A PRIMER OF GIS

A PRIMER OF **GIS**

FUNDAMENTAL GEOGRAPHIC
AND CARTOGRAPHIC CONCEPTS

Francis Harvey

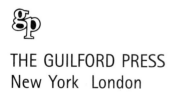

THE GUILFORD PRESS
New York London

© 2008 The Guilford Press
A Division of Guilford Publications, Inc.
72 Spring Street, New York, NY 10012
www.guilford.com

Printed in the United States of America

This book is printed on acid-free paper.

Last digit is print number: 9 8 7 6 5 4 3 2 1

Library of Congress Cataloging-in-Publication Data

Harvey, Francis (Francis James)
  A primer of GIS : fundamental geographic and cartographic
  concepts / Francis Harvey.
      p. cm.
  Includes bibliographical references and index.
  ISBN-13: 978-1-59385-565-9 (pbk. : alk. paper)
  ISBN-10: 1-59385-565-6 (pbk. : alk. paper)
  ISBN-13: 978-1-59385-566-6 (hardcover: alk. paper)
  ISBN-10: 1-59385-566-4 (hardcover : alk. paper)
  1. Geographic information systems.   2. Cartography. I. Title.
G70.212.H38 2008
910.285–dc22

                                                        2007050932

About the Author

Francis Harvey is Associate Professor of Geography at the University of Minnesota. He has also worked at the University of Kentucky and at a variety of academic and professional positions in Germany, Switzerland, and the United Kingdom. He has taught GIS courses in other academic and professional programs around the world. His research is wide ranging, with a current focus on governance of land and spatial data infrastructures. He received his doctorate in 1996 from the University of Washington for research on GIS overlay.

Preface

The idea behind this book is simple: to put in the hands of people interested in geographic information systems (GIS), geographic information science, and geospatial science and engineering a book that provides a broad preparation for later work with geographic information, *regardless of background*. Accordingly, this book explains, with a pragmatic approach, the concepts and practices of geographic information that underpin GIS. It covers what and how geographic information represents, analyzes, and communicates about human and environmental activities and events on our planet.

In order to serve a broad array of readers, this book has four parts that, read sequentially, build on each other to offer a successively deeper understanding of GIS. Part I introduces the most basic concepts of cartography and GIS; Part II goes into more detail to offer an overview of the fundamentals of cartography and GIS; Part III focuses on specific techniques and practices; Part IV looks at geographic information analysis and sketches out some of the exciting new GIS developments. Each part, or individual chapters, can be read separately or together with other parts or chapters for courses, seminars, training, and workshops to learn about specific conceptual or practical issues.

Most readers should start with the first chapter to make sure they understand the key concepts of geographic representation and cartographic representation. The other parts and chapters can be read as an instructor suggests or as fits your needs best. Given the breadth of GIS and the diversity of people reading this book, and its modular structure, some parts of the book repeat other parts: the repeated material may be well known to some readers, but useful to other readers who need different explanations.

The access point sidebars in some chapters provide detailed practical examples of how people use geographic information; example sidebars focus on relevant aspects of examples; exercises allow you to apply theories and concepts to learn skills; in-depth sidebars offer practically oriented detailed discussion of theories and concepts. To assist your reading and learning, you will also find Internet links at the end of each chapter to help you find examples that are relevant to your interests or learning needs.

This book provides a conceptual introduction to GIS without requiring the use of GIS software. Through practical examples and exercises, regardless of your educational background or interests, you will find in this book detailed introductions to the theories, concepts, and skills you will need to prepare for working with GIS.

Acknowledgments

Many people are explicitly connected to the writing of this book; many more implicitly. Above all I am happy to thank Martin Galanda for discussions in conjunction with the GEOG 1502 course we teach at the University of Minnesota. My other colleagues in the Department of Geography have been helpful on many occasions, particularly Mark Lindberg, Jonathan Schroeder, and Julia Rauchfuss, who were a great aid in preparing many of the figures. Over the years, numerous discussions with colleagues from the University Consortium of Geographic Information Science have led to the refinement of many of the concepts and skills I cover in this book. Colleagues and friends from around the world have also helped me out in various ways. I thank the following people for discussions and contributions: Adam Iwaniak, Marek Baranowski, Brett Black, Omair Chaudhry, Nathan Clough, Jason Dykes, Dietmar Grünreich, Peter Fisher, Randy Johnson, Chris Lloyd, William Mackaness, Robert McMaster, Lori Napoleon, Annamaria Orla-Bukowska, and Nick Tate. I most of all want to thank Alicja Piasecka and Anna Piasecka for their support during the many hours spent writing and revising this book.

Without their help I could not have written this book; any misinterpretations or errors in the presentations or translations remain my sole responsibility.

Contents

Part I

Communication and Geographic Understanding

Chapter 1

Goals of Cartography and GI: Representation and Communication

Many of our representations and communications about things and events around us, in history, even in the future, rely on geography and cartography. Usually we simply forget how commonplace maps and geographic information are, so maybe you have never given it much thought. Nevertheless, maps and geographic information are essential to how we know the world. The endless complexity of the world around us presents us with a multitude of choices about what to represent and how to represent that complexity in the form of maps and as geographic information.

Right now, take a look out a window. If you have a map of the same area, also look at that map. Compare your view to the map or to a map you remember of the place you are looking at. They are obviously different. Try to make a list of the differences. What is different between the view and the map? There are many, many differences: trees, buildings, or sidewalks may be missing on the map, the color of the road on the map may set it apart from other roads, the connections between roads may be much plainer on the map than what you can see. How and why geographic information and maps are different from our experiences and observations are important questions that this book will help you understand. Geographic information systems (GIS) involve many issues and choices and you are just at the beginning of the book; this chapter and the following two chapters provide a general introduction, with more detail to come in the other chapters of the book. As you read this and look at the map and out the window at the same area, you can start thinking about how your observations and perceptions of things outside are different from the map: some things are missing, some things are simplified, and some things are exaggerated on a map. Geographic information and maps are representations that follow a number of principles and conventions that help deal with the complexity of the world

FIGURE 1.1. Three modern maps showing geology, landforms, and political boundaries; each relies on different forms of geographic representation and cartographic representation to communicate particular meanings. Concepts and conventions of color and scale are crucial to assuring that their intended audiences understand each map.

and guide choices that lead to clear communication. Should the map include sidewalks? Will the geographic information describe the height of buildings? Are trees distinguished by species? These choices also will determine the way locations on the spherical earth are transformed to a two-dimensional plane, the types of colors and symbols to use, and the types of questions that people will turn to the map or geographic information to help find answers for.

Consider two other examples that highlight the different types of representation used in maps and geographic information (i.e., the data stored on a computer that contains information for making maps or conducting analysis) and point to some of the principles and conventions that guide mapping choices. First are maps of continents or subcontinents. You may never actually have seen the entire United States, all of Europe, or all of southern Asia in person, but you know something about how they are geographically organized through maps. Second, consider maps you use to help you get around the place where you live. You may know the way to go when you travel to work or school partly from descriptions prepared with the help of geographers and maps made by cartographers. Starting with these two examples, if you pause to think about the many different uses and roles of geography and cartography in the last 500 years (an arbitrary period), starting with the European period of exploration and colonization, we can conclude that geographers and cartographers have helped people to understand, navigate, control, and govern most of our world for millennia. Your world and the whole world would be much different without geography and cartography. We rely on these representations and the principles and conventions behind them to make sense out of the world in many different ways—sometimes geographic information and maps may be the only way to know something, other times they are important complements to other things we know or can ask. *Principles* are standard procedures that people in a field follow—for example, when a cartographer chooses a projection to make a map. *Conventions* are uses or procedures agreed upon by experts, but usually they have become common knowledge—for example, that north is the direction oriented at the top of a map. Sometimes we are sure about how things are geographically organized, but sometimes we may be less certain. We probably know where the city we live in lies in relationship to the coastline, but we may be less sure about

FIGURE 1.2. Three thematic maps from the 19th century that demonstrate different geographic representations and cartographic representations.

From *www.davidrumsey.com*. Reprinted by permission of David Rumsey.

whether New York or Boston is further to the east. A good representation takes these issues into account to assure that its readers or users find the representation helpful in communication.

Modern geography and cartography share many principles and conventions that form a symbiotic relationship, which make up an important basis for the geographic representation of the world in other scientific and professional fields. We define them in this book as follows. *Geography* analyzes and explains human and environmental phenomena and processes taking place on the earth's surface, thereby improving our understanding of the world. *Cartography* develops the theories, concepts, and skills for describing and visualizing the things and events or patterns and processes from geography and communicating this understanding. In this book things refer to elements of the world that are static, either by their nature or by definition. Events refer to selected moments in a process. Both are representations involving our innate cognitive capabilities and culturally and socially influenced knowledge of the world. What geography analyzes and explains, cartography communicates visually. Geography and cartography are dynamic subjects that involve a broad set of theories, concepts, and skills that undergo constant development and refinement as knowledge, culture, and technology change. Because of their usefulness, geography and cartography are parts of many other human activities and disciplines. Biologists, geneticists, architects, planners, advertisers, soldiers, and doctors are just a few of the scientists and professionals who use geography and cartography. However because geography and cartography are so commonplace, they are often easy to overlook. If you want to understand how to use and communicate better with maps, then you need to examine them closely and understand how and why geographic information and maps are different from what you see and observe. With a greater understanding of geography's and cartography's principles, conventions, and underlying basic concepts, you will be able to work better in any field.

For most people, maps are the most common way to learn about geography. But geographic information is very significant and continues to gain in importance. Geography and cartography have always been interdisciplinary fields. Many other disciplines and fields of human endeavor have drawn on their knowledge and skills and continue to do so. Recent technological innovations further broaden possibilities for people to make measurements of geographic things and events, operate and transform these measurements, and represent the measurements as information and maps. They produce geographic information, which is very easy to copy between computers, but often very hard to get out of the hands of the people and organizations who are responsible for that geographic information. Certainly, the circle of people working with concepts from geography and cartography has grown tremendously in the last 20 years. This has much to do with the increased availability of computers and programs for working with digital geographic information. That term sounds simple, but turns out to be highly complex. You might want to think about geographic information as you would about oxygen: you can't necessarily see it, but its presence has positive effects for people. Maps rely on geographic information. Geographic information is, of course, very different from maps in many ways. One of the most fundamental differences is that geographic information is very, very easy to change, whereas maps, if changed, are usually somehow destroyed. This means that geographic information can be used many times, which gives it a great advantage over maps.

Indeed, many geographers and cartographers would claim that geographic information makes geography and cartography more accessible than ever before. Farmers use global positioning system (GPS) technologies and satellite images to help disperse fertilizers and pesticides more accurately, safely, and economically. Fire departments route fire trucks to their destinations based on analysis of road networks and real-time traffic information. You may even have had the chance to experience these changes or to use GPS when navigating a boat, planning a trip, or driving a car. Many cars now come equipped with satellite navigation systems that rely on dashboard map displays to help drivers find their way. GIS is used also in many research facilities and offices to help analyze and manage resources. Improved geographic and cartographic technology has played a key part in important economic developments not only now, but in the past as well. The astrolabe used by navigators in the Middle Ages changed the way locations were determined and mapped; exploration consequently became more accurate and

FIGURE 1.3. Geographic information and maps show things and events from built and natural environments. The primary difference is change. Things are static for the observer, whereas events record selected moments of a process.

TABLE 1.1. Some Common Things and Their Representations

	Geographic Representation (Basic)	Cartographic Representation
Stream	Line	Color blue
Road	Line (usually)	Color black or red
Forest	Polygon	Color green
Industry	Polygon	Color gray
County or district	Polygon	Dashed boundary line
Well	Point	Circle with cross
Land parcel	Polygon	Thin black boundary line
House	Polygon	Thick black outline
Lake	Polygon	Color blue
Park	Polygon	Color green
Sand dunes	Polygon	Black dots on sand-colored background

safer. Offset printing, introduced in the late 19th century, made it possible to produce series of maps by using combinations of different plates; maps became commonplace in books, magazines, and newspapers. The most significant current geographic and cartographic innovations arise from the computer and the development of information technology for processing data during the last 40 years. The fields of geography and cartography entered an unparalleled period of symbiosis with the introduction of information technology for processing geographic information. This symbiosis resulted in a new field called geographic information systems (GIS), which, since the 1960s work by Roger Tomlinson, Edgar Horwood, William Warntz, and many others has grown into a major information technology field and a science.

People from many academic backgrounds correctly point out that the relationship between geography and cartography has changed and continues to change as a result of technological change; sometimes they even question the future of cartography because of GIS. Now, some people assume, computers can do all cartography. However, it is apparent that many of the key geographic and cartographic concepts established over thousands of years remain important. In fact, one could claim that these fields are really not changing conceptually, but only in degrees. As information technology becomes commonplace, many more people are now able to do things without the years of training that only cartographers and geographers previously had. Of course, because of all the people now doing work with geography and cartography on computers, one could also argue that the underlying concepts and skills of geography and cartography have become more relevant. Both are certainly true; however, without understanding of the concepts and skills, the best intentions can easily go wrong. Obviously, professionals always need to produce the highest quality maps and always benefit from better understanding of the concepts and skills—regardless of what and how much information technology is capable of doing.

Representation and Communication

In this book, you will learn about both the old concepts and the new concepts within cartography and geography. You will find that the old and the new concepts of geography merge with new information technologies in representation and communication, the two essential activities of geography and cartography. In this book "representation" refers to the active process of observing the world and symbolizing those observations to make meaning. "Communication" means the process of presenting these representations by some people and the viewing, or reading, of those representations by other people. Geographers and cartographers are always involved in communication, for even if it is not their immediate goal, maps and geographic information are always made to share information and knowledge about the world. A "geographic representation" is the specific process of abstracting observations of the world into things or events, often resulting in a model. "Things" are the results of activities, measured properties of objects or features, and distinct characteristics about people, places, or situations. "Events" are records of processes—for example, the movement of cars and trucks, the flow of water, the melting of ice, or the spread of a disease. A "cartographic representation" involves the process of symbolizing the geographic representation. Successfully communicating information about things and events requires you to know something about geographic representation and cartographic representation. These two concepts include color, symbology, modeling, projections, and, now with GIS, spatial database queries and attribute types (all covered in later chapters).

This book considers representation and communication as related and fundamental topics in geography and cartography. A peculiar geographic fascination is common among people working with GIS, whether they work for a utility company, a county government, a university administration, or a corporate marketing department: How can the infinite complexity of the earth's surface and related processes be reliably represented? This seemingly abstract question touches on the key issues these people have become aware of through their education, training, and work experiences. They must decide how to represent selected things as patterns that show important elements and processes in relationship to the places where they take place. Figure 1.4 shows simplistically a few basic choices and the different ways events can be represented either by highlighting the process or by translating the site of the process into a pattern. How representation is chosen also must consider the context of the intended communication, particularly the reader's/user's knowledge and background: How well does the application or map correspond to what the readers/users know or could know? Are data available to provide that information? How long would it take to acquire new data? The issues include many specific questions—for example, Is it sufficient to show trees as points where their trunks are located or as areas that show the reach of the foliage?

The answers to the question of representation usually come back to

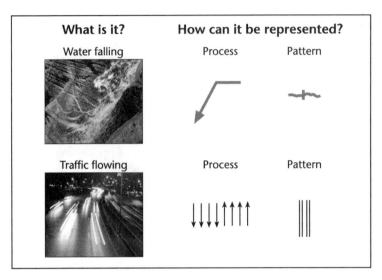

FIGURE 1.4. Events can be represented in maps as both processes and patterns.

choices and quality. There are many choices: How much detail is needed for the geographic information or map? Will a poster-size paper map be needed for the detail or does it have to fit on a small handout? How accurate should it be? How big should (or can) it be? If this is a computer application, will the data be available on a CD-ROM or a DVD, or will it be downloadable over the Internet? How big is the screen display? The type of communication and the background of potential users also need to be taken into consideration: Will specialists use the map or application? How much knowledge do they have about the area? What is/are the specialists' purpose or purposes? How much contextual information is required? How abstract can the representation be? How reliable must the representation be? Each decision influences quality in complex ways. If the map needs to fit on a small piece of paper, but the area of an entire state or province needs to be shown, it will be very difficult to show a great deal of detail.

Issues related to wise choices and quality come back to perennial issues for geographic representation. The space of the earth's surface is limited, but because all geographic information is an abstraction with no limit to the number of choices we may make in presenting it, geography's potential representations are unlimited. The space of the earth's surface shows itself in peculiar characteristics in every representation. How close objects are on paper or on the screen depends on the relationship between the size of the representation and the actual area on the ground. This is what geographers refer to as "scale." Scale is a crucial component of geographic representation and cartographic communication. Of course, people think objects closer together on a map are more related to each other than objects far apart, but if you consider the scale of the representation, even-close objects may actually be very distant from one another. The issue of scale is a particularly central concept for all geographic information because of the ways it allows and restricts representation, the communication of relationships between things, and the interpretation of geographic associations.

The Power of Maps

Successful answers to these questions and attention to the decisions made in representing geography are what gives geographic information power and makes maps powerful, to borrow from Denis Wood's thoughtful writing about maps. A map or geographic information application is selective and greatly limited, but it remains a key means of understanding and analyzing the world. This power is very attractive and lucrative; its misuse and abuse lend support to many ill-conceived projects.

Maps are powerful for a number of reasons. But perhaps the most elementary reason is that they offer an authoritative representation of things and events in the world that we cannot otherwise experience in a single moment. Most maps, even the most mundane kind of map—for example, one showing temperatures across North America—show us things, events, and relationships that you could never experience yourself in a similar complete but quickly grasped form. You can read a book, look at a photograph, watch a film, check things out on the Web, but a successful map easily and quietly combines much detail into a synoptic whole.

FIGURE 1.5. The power of maps depends on the currency of the map. In 1844, when this map was prepared for the U.S. State Department, it played an important role in helping people understand the Texas conflict.

From *www.davidrumsey.com*. Reprinted by permission of David Rumsey.

FIGURE 1.6. The power of maps is significant for associating organizations with a nation or region. This sign for the Polish Tourist Association uses an iconic representation of Poland's national boundaries.

Perhaps the second elementary reason why maps are so powerful is that they represent something beyond our own limited experience: other people and places you may never see in person, other things or events that we may never know about otherwise. They became a key source of information about people and places we can't experience because of distance or because of complexity. Maps become a primary source of information for many things since we often cannot verify what they tell us. Is the Eiffel Tower located at the center of Paris? Unless you are in Paris or will be shortly that cannot be determined except by using a map.

The power of maps comes through their ability to create representations of the world that most people won't question because they lack the direct experience of the people or places, things, or events to evaluate the representations. It is very hard to know that a representation implicitly makes a threat out of a neighbor, errs in creating symbols that mask important details, or explicitly shows a part of the world in a biased manner. Using red to show the country of one's enemy awakens a sense of menace because most people associate the color red with danger. Showing a country in green has the opposite impact. Because they follow frameworks and conventions that we have become used to, slight distortions are easily veiled and become undistinguishable.

Maps are often misused and have become important tools for propaganda and advertising (see Chapter 11). Extreme examples clearly show abuse of cartographic integrity, but you also need to be wary of more common and subtle misuse of map power to create biased representations.

Types of Maps

Three of the most common types of maps are thematic, topographic, and cadastral. There are many ways to develop typologies of maps, but these three types seem to distinguish both how and why maps are used. *Thematic maps* are the most common: they show specific topics and their geographic relationships and distributions. Thematic maps show us the weather forecast, election results, poverty, soil types, and the spread of a virus. *Topographic maps*—from the United States Geological Survey (USGS), for example—show the physical characteristics of land in an area and the built changes in the landscape. *Cadastral maps* show how land is divided into real property, and sometimes the kinds of built improvements. How each type of map is made with geographic information is a question that you will be able to answer generally at the end of Part 1. You can find out about the specific

FIGURE 1.7. Maps are generally distinguished by scale and whether they are thematic or topographic in nature. Each type shown here is characterized by different geographic representation and cartographic representation choices.

concepts and skills in Parts 2 and 3. The types shown here are an arbitrary selection intended to show how types of maps vary at different scales.

Mental Maps

Many people find that mental maps are a great way to start thinking about how maps represent and communicate about the world. Thematic, topographic, and cadastral maps are useful for communication because they follow known and accepted conventions, but they often have little in common with our day-to-day experiences. Mental maps are much stronger on this point, but suffer from weaknesses as a reliably understood means of communication. Mental maps communicate what an individual knows and can draw about some aspect and part of the world. A mental map represents particular geographic relationships based on the experience of an individual. A mental map communicates those relationships from the perceptions of one or sometimes a small group of people, but often can be difficult to understand without some form of description or use of standardized cartographic representations.

Based on human perception and behavior, Kevin Lynch developed mental maps in the 1950s as a planning technique for understanding how a city was legible. "Legible," for Lynch, meant how well the structure and organization of a city helps supports people's lives by being easily understood and requiring a minimum of effort. Using systematized graphic elements, Lynch cartographically represented people's mental maps of the city to show how they perceived and moved about the city. Mental maps are often used to help planners gain a better understanding of what features in the city need improvement or change. Many researchers have gone on to use mental maps along these lines to assess gender, race, or age differences in urban experiences and life. It is important to remember that mental maps generally lack a consistent scale or set of symbols. Because they are usually purpose-oriented and based on the selective memory and knowledge of one person or group, they are incomplete by nature and often hard for others to use. For example, in Figure 1.8, the dashed lines connecting the person's home neighborhood to downtown could indicate any distance; the readers of the map can only know how great or small a distance if they know the drawer or the area.

Geography and Cartography in Harmony

To successfully use GIS and make informative maps, geographic representation and cartographic communication must work together. Before getting into the details later in the book, let's look at the how geographers and cartographers usually understand and represent the world. You may already know how your field or profession makes geographic information and maps. However, your work with maps and geographic information may greatly benefit from thinking about the conventions in your field or profession and the

FIGURE 1.8. Mental map showing the route from to the center of town. Note how the cartographic representation collapses space to remove or highlight significant geographic detail for this person.

From *www.subk.net/mapsindex.html#maps.* Reprinted by permission of Lori Napoleon.

assumptions that go along with them. Much of this discussion can easily become part of complex philosophical discussions about existence, knowledge, and representation, but we will skip that for now to sketch out more pragmatically what many geographers and cartographers think about when figuring out how to understand, analyze, and represent the world around us. You need to get a basic idea of how maps and geographic information require a multifaceted framework with many conventions. Understanding the framework and conventions will be the basis for considering concepts and skills required for making and using maps and geographic information.

Things and Events

If something can be represented geographically, it is either a "thing" or an "event." These are the terms most people also commonly call what you find represented on maps. You'll see later some other terms, such as "feature" and "object," that will help you think about the possibilities and limits of maps and geographic information. Right now, however, let's consider how things and events can be represented in geographic information or maps. *Purpose* is an important factor for guiding the choices made when making and using maps and geographic information. For an example, assume you want to make a map of places where you live where traffic jams occur. You need to show the location of the traffic jams and the roads they occur on. The map should also show the location of attractions and important land-

FIGURE 1.9. Image showing the results of stream modeling combining a geographic representation of events and their translation into the cartographic representation of things.

From *http://wwwbrr.cr.usgs.gov/projects/SW_Math_mod/ OpModels/MD_SWMS/Presentations/ Flood%20Inundation_files/v3_document.htm*

marks to help people navigate who are unfamiliar with the roads and area. Showing the things that are traffic jams is helpful for people visiting town, but it may not be enough if you want to help someone coming into town at 5:00 P.M. find the best way to avoid known traffic jams to the best restaurant on Main Street downtown. For this map, you need to show not only *where* traffic jams occur but *when* they occur so that your visitor knows which roads to take and which roads to avoid. It would be better for the purpose of guiding your visitor to just show the traffic-jam events that can hinder his or her trip to the restaurant. Showing too much information about attractions and landmarks will probably be unnecessary for your visitor to find his or her way.

Remember that things are static geographic information or map patterns even though they can refer to a process such as too many vehicles traveling at the same time. Animation techniques and dynamic GIS offer some interesting possibilities to create dynamic cartographic visualizations that show temporal changes. Geographic information analysis and geostatistics provides a number of techniques for reliably representing processes (see Chapter 13).

Abstraction and Reliability

Part of what makes representation difficult is that it simultaneously abstracts while it attempts to assure reliability. *Abstraction* reduces complexity, or simplifies, to highlight essential things, events, and relationships. *Reliability* is the characteristic of a representation that refers to its dependability.

Representing things and events is complex because of the very nature of abstraction. The world is theoretically infinitely complex: it is not possible

FIGURE 1.10. The map on the left shows current road status (events), the map on the right shows highway projects (things) that can affect traffic. Each map uses a different cartographic representation as well.

From *www.dot.state.mn.us/tmc/trafficinfo/map/refreshmap.html* and *www.511mn.org/ default.asp?area=TC_Metro*. Reprinted by permission of Mn/DOT.

for anyone to make a map of everything in the world; nor is it possible to make a map of any area showing everything at that place, even at just one time. Every map is an abstraction that focuses on a selection of things and events from that place according to the purpose of the map. A highway map emphasizes roads and represents buildings, rivers, and towers in the landscape as context for using the map to navigate. A map of forest fires shows the location of fires and represents the slope and exposure of the hills or mountains with contour lines.

Geographers and cartographers want these representations to be reliable for the purpose they are intended for. A highway map is good for driving in a car; it is less useful for riding a bicycle and of little use for planning a trail hike in a state park. The maps of forest fires in a national park are good for understanding where forest fires occurred, but may be less helpful to determine why the fires occur and where they may occur in the future, and not of much use at all when planning a hike.

The reliability of a map or geographic information depends greatly on the choices made in abstracting things and events to the static patterns shown on a map or the information stored on a computer. Following Nick Chrisman, these choices are part of a framework encompassing measurements that record aspects of geographic things and events; representation of these measurements as geographic information to indicate geographic things, events, and associations; operations on these measurements to produce more measurements; and transformations of the representations to other frameworks. The integrity of the process of representation makes for reliable maps and geographic information. You will see later how this merges together into what can be called geographic information representation (Chapters 3 and 9).

Space, Things, Events, and Associations

All geographic measurements start out with observations of things and events. Things and events may seem at times to be independent of each other, but geographic relationships bind them together in associations. How you approach them is not only a matter of geography or cartography, but also a matter of a field's or a discipline's conventions. A geomorphologist thinks quite differently about streambeds than a limnologist. A city planner thinks that street centerlines are good for zoning boundaries; a geodesist may differ. You may have your own examples from your field. Many GIS scientists think about these associations in terms of a predicate calculus that can be manipulated to model the situation and develop stable descriptions. Most people using GIS are glad when they get the results of this science, but they pragmatically focus on working with what they know and improving that knowledge and their abilities. This latter point is the focus of this book, although finding out about the underlying science is important to learn about too.

Geographic information science (GIS), the field concerned with the underlying theories and concepts of geographic information, is pertinent when you learn about geographic and cartographic concepts for maps and geographic information. You might already be familiar with the terms "spatial" and "geospatial," which refer to properties that take place in space, especially activities on the earth. These terms refer to understandings of the world slightly different from geography is with its interest in places and spaces. These two terms suggest that the work described with these terms is usually done for purposes outside of a traditional understanding of geography. In any case, these terms, along with "geographic," are for this book's purposes synonymous. However, you should be aware that the terms used in this book can vary in meaning among disciplines and settings.

The underlying disciplinary concepts of space, relationships, and associations also can vary greatly. Space is a continuous area. Things and events in space can be related or associated. *Related* means that the things and events are connected in terms of distance. *Associated* means that things and events occur together, without any intervening distance. This book adopts a pragmatic perspective regarding disciplinary concepts, which is related to an empirical and contextual understanding that things and events mean what they do because of who is creating the meaning and in which context.

Frameworks and Conventions

Many people look at maps dubiously. They say maps don't make any sense; they don't match what they see; they are far too complex. Many other people almost feel lost going somewhere without a map. Why is that? There are certainly many personal and subjective reasons involved, but I want to suggest that a great deal of the troubles using maps can be helped by knowing the frameworks and conventions of cartography established over the last 500 years by Western civilization (and before then by other civilizations). These frameworks and conventions are crucial to understanding what maps and geographic information show us, how we understand them, and how you make maps and geographic information.

You can think of the cartographic frameworks as a set of normative and acceptable ways for showing things, events, space, relationships, and associations. Conventions are the actual ways of representing and communicating in maps and geographic information. They both vary as greatly as any fashion.

The elements of our cartographic framework and conventions are very profound and important in ensuring that maps and geographic information make sense to people. Obvious examples are the almost universal north orientation of maps and the depiction of water with a shade of blue. It's not too hard to think of other examples. But if you can look at maps from different countries, you'll start to recognize certain dissimilarities in the map symbols of each country. Even countries similar in cultural backgrounds can use symbols very differently. States in U.S. highway maps are usually drawn in a pastel color, but European maps of all of Europe choose more vibrant colors. U.S. maps generally don't show individual mountain ranges, but European maps generally show mountains with shading.

The frameworks and conventions can also serve particular professions and ideologies. Maps for specific uses, professions, and disciplines usually follow a number of conventions for simplifying complex concepts that make it very hard for untrained persons to make good maps. If you aren't familiar

FIGURE 1.11. Even simply flipping a map of the world so that Antarctica and Australia are at the top of the map can prove confusing because of the different cartographic representation.

FIGURE 1.12. A combination of maps and signs located on city streets helps people with finding their way in many cities helping people understand the cartographic representation of the city and relate it to their experiences.

with the frameworks and conventions, then maps are very complex and hard to read. That's why in places where almost all people need to use a map (at some point), that map is usually very simple and shows a specially highlighted symbol where you (and the map) are. Geographic information and most maps lack these. They are very frustrating to use without understanding the framework and conventions of maps and geographic information. Of course, we need also to consider the cultural biases and ideologies of maps. Many a European map of newly discovered areas conveniently erased most traces of aboriginal inhabitation; later, many Western maps showed religiously significant sites with blatant disregard. Because maps and geographic information are often the only sources of detailed geographic information, many people accept them as the best indications, even though biases may be great and the cartographic communication only partially works.

Quality and Choices

A common ward to describe the reliability and integrity of maps and geographic information is "quality." Most people consider a highway map to have good quality if they can use it to find their way easily. It may not be good to find the way to a hotel in the center of the city, but it is very useful for finding the way from Los Angeles to Portland. Simply said, good-quality geographic information or maps are useful for the purpose we create or intend to use them for. Quality usually means reliability, but in regard to maps it often also means that the map maintains integrity regarding the world and fits the use we intend the map for. We could turn to some highway maps to figure out the size of towns in a state, but most highway maps will not help determine the extents of the towns. The concept *fitness for use* helps get a grip on the slippery concept of quality. What this means is that we need to know the intended purposes of a map before we decide what degree of quality the map has for certain uses. You can read more about fitness for use in Chapter 11.

For now let's let quality mean "fitness for use," which is the ability of the final product to support a particular task. Because of the endless number of choices people face in cartography and geography, the quality of a map or geographic information is related to the purpose, data sources, and data processing. A highway map follows geographic representation and cartographic representation choices that enhance its suitability for driving. While it is obvious that a map showing all of the United States, Europe, or India is not well suited for finding the nearest restaurant from an office, house, or hotel, many maps and much geographic information are intended for multiple uses, which will give them both strengths and weaknesses. Geographic information may be collected and prepared for purposes that invite many different uses, but it may in fact be ill-suited for particular purposes. In the end, most maps are quite limited: a highway map is not very helpful for hikers (except for helping them get to where they want to hike). Geographic information can be handled more flexibly, but this does not remove significant limits. Well-known errors have occurred when people creating road databases included ferry routes as connections for national roads and then entered them as roads in a car navigation system. More than one driver has taken a surprise bath as a result.

Some of the many choices for map and geographic information quality are fundamental. All are topics that this book examines in depth. As you get a better grip on these choices, you will better understand the quality of maps and geographic information and also how to create better maps and geographic information. Below is a list of important topics and the chapters in which they are covered.

Projections: How geographic locations on the round earth are shown on a flat map or coordinate system is one of the biggest choices affecting quality (Chapter 4).

Coordinate systems: Related to a projection, a coordinate system is especially pertinent for geographic information, which can easily be combined with other geographic information when it is in the same projection and coordinate system (Chapter 5).

Symbols: How things and events are communicated is certainly one of the biggest choices affecting quality. For most people using maps, it is the most important, because if people can't make sense out of the map, how can anyone ever judge the quality (Chapter 10).

Geographic representation: Deciding how to show things and events is crucial to whether a road can be modeled with different lanes of traffic and sidewalks or only in terms of traffic flow.

Cartographic representation: If the geographic representation provides the information, the cartographic representation can support various representations, contingent on a number of parameters, notably scale.

Conventions and Quality: An Example

You can probably think of a number of times when a map wasn't as helpful as you wished it would be. That may well have been because of quality and choice issues. Before moving to more specifics of communication and representation in the next chapters, you can follow this example to see how frameworks and conventions in relationship to quality and choices can lead to a less-than-useful map. This is a very big problem for companies whose business depends on maps, so they put great effort into making maps understandable, but have to make some important choices that may greatly limit the quality of their map for some groups of people.

People who travel often have to use rental cars and depend on their maps. If you arrive in a country with a strong mapping tradition and rent a car at the city airport, you may receive a very good road map of all of the country; this map may be perfect for finding the smallest town that your friend's ancestors originate from, but is probably much less perfect for finding the way to your hotel. But let's assume that you have been given a pretty good map of the city, one that even shows hotels. It's no problem to find the hotel you're going to . . . but wait. The center of the city is on the map, the hotel is there, but where's the airport? What do you do now? Where is the airport? What's the road to the city

**Oak Ridge–Knoxville Area
Route Map**

FIGURE 1.13. A directional map can be useful for finding some directions, but its geographic representation will fall short in supporting other uses, such as finding the way from the airport to a hotel.

From *www.y12.doe.gov/library/maps/OR_KnoxMap.gif*

called? The framework for such maps is suited for people who know the city and speak some local language and who are looking to find their way around, not for visiting businesspeople or tourists. In this example, not at all uncommon, you can summarize that the conventions for the local maps support the general orientation of people who are familiar with the conventions of these maps and the culture, not the specific conventions and definitely not the purpose of a visitor from another country.

Distinguishing Geographic Information from Maps

Maps remain important, but more and more maps are produced with geographic information. Some people now even suggest that most maps are simply interfaces to geographic information databases. Several years ago, separating geographic information from maps would have been complicated. Maps, following the International Cartographic Association, are science and art. Geographic information was interpreted or symbolized data. It's simpler now. In this book "maps" are a form of output of geographic information. Maps are truly the most common form of output and have been essential to our understanding of the world for millennia. Maps can be drawn by hand, and constructed by hand, but nowadays are mostly prepared using geographic information.

The computerization of cartography changes the possibilities you have for working with the underlying geographic information. Geographic infor-

FIGURE 1.14. Vector line data using a cartographic representation that distinguishes road types.

FIGURE 1.15. The same features from Figure 1.14 shown without any cartographic representation.

mation is usually presented as maps, but tables, figures, and hybrid output forms are also legitimate output forms. Geographic information is what is used in GIS. Data is what information is before it is used and makes sense to the persons creating or using the geographic information or map. Earlier in this chapter, I compared geographic information to oxygen. Now, starting there, you can think of one of the effects of information: information has at least the potential of having an effect. Data may sit in an archive for years and years—never having an effect until someone looks through the data, makes sense out of it, and "converts" it to information.

Geographic information is not data. Data can become information, or may have been information, but it is only the raw recording of measurements used for creating information. To become information, it must be put into relationship with a purpose (or purposes) and (potential) use. Data can simultaneously be information and data for two or more people, if one person uses it unchanged as information and another uses it as the basis for creating information. In other words, one person's data is another person's information. How do we know? Apply the sense test. If what you see, regardless of its form, makes sense, it is information because sense, or meaning, only comes if the thing you see has an effect; if it doesn't make sense, it is just data. Of course, this only means it makes sense for you. Making the same sense for others is a much harder, but more important, test.

Summary

Geographic information and maps are ways of representing what people see and observe. Things and events should be distinguished. *Things* refer to static representations of something in the world; *events* refer to dynamic changes. Because of the complexity of the world, even small aspects of it, and the challenges of representing the world, a number of choices are brought together in geographic representations and cartographic representations. Because of the large number of choices, cultures, fields of science, and professions rely on conventions and frameworks. *Conventions* are often unstated guidelines for representation. *Frameworks* are rules and procedures for dealing with the complexity of choices. Geographic information and maps have a great deal of power as a result of making portions of the world understandable. The quality and reliability of geographic information and maps depends on how well they fit the purpose, or "fitness for use."

In-Depth Globes

For many people globes seem to be ideal cartographic representations of the earth's geography. They are certainly attractive, but for a number of reasons they are limited in their use and suitability for most maps and geographic information. They remain, however, the best reference for understanding the earth's three-dimensional shape and for conceptualizing latitude and longitude.

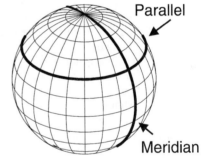

Globes indicate lines of latitude and longitude, also called respectively parallels and meridians.

Making a Globe

A globe ends up round, but it is printed on a flat piece of paper just like any other map. The map is divided, and later cut, into what is called a "gore." When the paper is glued to the round base, these strips fit together, resulting in an uninterrupted sphere. This technique has been used for several hundred years.

Reading a Globe

Locations on a globe can be found by latitude and longitude. *Latitude* is a degree measure used to indicate the relative distance from the equator. *Longitude* is another degree measure that has a fixed origin, nowadays the Greenwich meridian, running through Greenwich, England, just outside London. Meridians run from pole to pole; parallels run from east to west at a constant latitude.

Waldseemüller globe gore from 1507. From the University of Minnesota James Ford Bell library. Reprinted by permission.

In-Depth Representation

An artist may represent a face, a room, or a landscape. A computer scientist may represent the same things and events as a data model or a database. A geographer may represent these things as objects and as relationships. A cartographer may represent them as features and attributes.

For some people, representations have dangerously replaced experience. On the other end of the spectrum, other people find that our enhanced technologies for representation make it possible to transcend past limitations. We can talk to people all over the world, they may point out, but others ask, For what good? This is not the place to engage these discussions, but before getting into more specifics about representation, it's good to have some notion of how the term *representation* is used.

All these forms of artistic, computer, geographic, and cartographic representation are important to how we make sense of the world around us, but our vocabulary is a bit weak for distinguishing between them. To help with this problem, this book always proceeds "representation" with an adjective to distinguish what kind of representation is meant—for example, *geographic representation*.

Review Questions

1. What distinguishes a *thing* from an *event*?
2. What is the geographic significance of the difference between a *thing* and an *event*?
3. How are things represented in cartography?
4. What influences the quality of GI or a map?
5. Is every map reliable enough for every use?
6. How do representation and communication relate to each other?

7. Why can't GI or a map show everything in any chosen area without distortion?
8. What role do conventions play in the creation of GI and maps?
9. Can GI or a map be understood just by itself?
10. What are the differences between GI and maps?
11. What makes maps so powerful?

Answers

1. What distinguishes a *thing* from an *event*?

   A *thing* is a geographic representation that is static and undynamic; an *event* is a process or part of a process characterized by a change in how it is being described and accounted for. This difference is significant for cartographic representation and cartographic communication. Things can be easily portrayed with cartographic symbols on paper or other media; events can be only portrayed as a series of "snapshots" or through animations.

2. What is the geographic significance of the difference between a *thing* and an *event*?

   The difference between a *thing* and an *event* is significant geographically and also for cartographic representation and cartographic communication. The geographic representation of things emphasizes consistent characteristics; events emphasize changes or processes.

3. How are things represented in cartography?

   Things can be easily portrayed with cartographic symbols on paper or other media; events can be only portrayed as a series of "snapshots" or through animations.

4. What influences the quality of GI or a map?

   The choices made that affect the reliability and integrity of the GI or map.

5. Is every map reliable enough for every use?

   No, maps are abstractions that focus on a selection of things and events from the endless complexity of the world.

6. How do representation and communication relate to each other?

   Representation is the basis for communication. It is impossible to have a cartographic representation of what is missing from a geographic representation.

7. Why can't GI or a map show everything in any chosen area without distortion?

   Maps must abstract, which leads to distortions. Some distortions are explicit, but many can be implicit.

8. What role do conventions play in the creation of GI and maps?

   Conventions are simply unstated rules and assumptions that people rely on to help with the geographic representation and cartographic representation of the world.

Environmental Monitoring in Central and Eastern Europe

Dr. Marek Baranowski has served as the director of the Warsaw, Poland, office of the United Nations Environmental Program, Global Resource Information Database (UNEP-GRID), since 1991. He is a geographer by background and has worked with GIS since 1973. This office is involved in many environmental projects in Central and Eastern Europe, which since 1989 have seen rapid change. Educational and learning resources prepared by UNEP/GRID-Warsaw have twice won (in 2001 and 2003) first-place awards from the Polish Ministry of Environment for outstanding achievements in environmental science and development. He and the office are also involved in several national and regional environmental monitoring projects, visualization for participatory planning, the Polish general geographic database, and EuroGlobalMap.

GI and cartography are central to these projects. The educational and learning resources use GIS to collect and prepare data; maps are central to the multimedia educational tools used in classrooms across Poland. The coordination of national and regional environmental monitoring involves GI and maps. Working on the Carpathian Environmental Outlook, which is connected to the Global Environmental Outlook, requires the coordination of information from the countries of Austria, Czechia, Hungary, Poland, Rumania, Serbia-Montenegro, Slovakia, and Ukraine. Because they need to show past changes, they must pay careful attention to the seasons, time, types, and resolution of remote sensing data used to detect changes. When we spoke, Dr. Baranowski told me that a previous project of the CORINE Land Cover project in Romania once had to be repeated in order to recollect all the data on the basis of new satellite images since so many changes had taken place between the start and the conclusion of work.

GI and cartography have limits for showing dynamic processes. According to Dr. Baranowski, detection of changes is the key way to show processes in this large area. Each human-environment interaction can not be detected and recorded in an area where over 15 million people live. Detecting changes becomes complicated because of different data collection issues. For example, satellite images collected in spring are different than images collected in the fall. Large errors in the determination of changes could result when using noncomparable images. Detailed technical specifications address these issues and other concerns. The challenges of dealing with human–environment interactions at this scale involve technology as much as organizations.

More information about UNEP/GRID-Warsaw is available at *www.gridw.pl*

9. Can GI or a map be understood just by itself?

It may seem this way, but that understanding arises because of the knowledge we have of conventions—for example, water is blue, north is usually at the top edge of the map, and so on.

10. What are the differences between GI and maps?

Maps are printed or displayed on a media which cannot be changed nor altered without altering the map. Maps cannot be altered except by destroy-

Carl Steinitz and Landscape Architecture

Landscape architects deserve special mention in the history of GIS. They were involved in developing many of GIS's first key applications including techniques based on traditional overlays of transparent thematic maps on a topographic base map. This overlay technique allowed the landscape architects to provide a situational map that could be shown with specific themes (parks, schools, ecotones, etc.) and with composites of the themes. This became a key part of GIS-based analysis. Ian McHarg was one of the key developers and promoters of the overlay-based technique for planning. Carl Steinitz, a contemporary of McHarg, stands out for his contributions to the development of overlay techniques, their application, and documenting the history of overlay techniques in landscape architecture.

Further Reading

Steinitz, C., P. Parker, and L. Jordan. (1986). Hand-Drawn Overlays: Their History and Prospective Uses. *Landscape Architecture, 66*(5) 444–455.

ing the map and reusing portions of it for other purposes. GI can be used over countless times, in different ways, to make different maps.

11. What makes maps so powerful?

Much that we know is known to us only through maps. The World, Asia, the United States, even an entire city are places of which we can only experience a fraction of. Maps put selected portions of things and events from the world into a comprehensible graphic format that communicates. Maps are powerful when they successfully communicate what we didn't know before.

Chapter Readings

Board, C. (1967). Maps as Models. In R. Charley & P. Haggett (Eds.), *Models in Geography* (pp. 671–726). London: Meuthen.

Buttenfield, B. P. (1997). Talking in the tree house: Communication and Representation in Cartography. *Cartographic Perspectives, 27,* 20–23.

Chrisman, N. R. (1997). *Exploring Geographic Information Systems.* New York: Wiley.

Dorling, D., & D. Fairbairn. (1997). *Mapping: Ways of Representing the World.* Edinburgh Gate, Harlow, UK: Addison Wesley Longman.

Gersmehl, P. J. (1985). The Data, the Reader, and the Innocent Bystander—A Parable for Map Users. *Professional Geographer, 37*(3), 329–334.

Kaiser, W. L., & D. Wood. (2001). *Seeing through Maps: The Power of Images to Shape Our World View.* Amherst, MA: ODT.

Monmonier, M. (1993). *Mapping It Out: Expository Cartography for the Humanities and Social Sciences.* Chicago: University of Chicago Press.

Monmonier, M. (1995). *Drawing the Line: Tales of Maps and Cartocontroversy.* New York: Holt.

Pickles, J. (2004). *A History of Spaces: Cartographic Reason, Mapping, and the Geo-Coded World*. New York: Routledge.

Thompson, M. (1987). *Maps for America*. Washington, DC: U.S. Department of the Interior, Geological Survey.

Wood, D. (1993). The Power of Maps. *Scientific American*, 88–93.

Web Resources

National Geographic's online Xpedition Hall offers a highly interactive introduction to geography and cartography. The Mental Mapper offers a good introduction to mental mapping and you can follow one of the lesson plans. See *www.nationalgeographic.com/xpeditions/hall/index.html?node=20*

Three hundred and twenty-one definitions of the word *map* are collected at J. II. Andrews's website: *www.usm.maine.edu/~maps/essays/andrews.htm*

The ICA website provides up-to-date information about cartographic research, teaching, and publishing around the world: *www.icaci.org/*

The University Consortium for Geographic Information Science (UCGIS) website offers information about some of the newest GI research: *www.ucgis.org*

For many, the USGS website is the first place to go to for information about mapping in the United States: *www.usgs.gov*

Exercises

1. Comparing Map Representations and Our Observations

Get a map for a well-known area. The map can be at any scale or of any type; it should show things that people are well familiar with. Start out by comparing what you remember of that area to how it is represented on the map. What is missing? What is simplified? What has been added or exaggerated? Compare your map and answers to these questions to a neighbor's. Are they the same or different lists? How could that depend on scale or the type of map?

2. Make Mental Maps and Discuss Them with a Classmate

Draw maps from memory of an area you and your neighbor are familiar with. When you are done, identify common elements with your neighbor or in small groups and write them down on the board in your classroom and discuss how well the mental maps help communication. Also discuss how the maps could be drawn differently and what cartographic or geographic choices the different map involves.

3. Choices and Scales

Examine the figure showing the three types of maps and different scales. What choices do you think were made to make each map? What is the scale of each map? How does the differences between map scales affect the way things are shown? How do the choices differ in relationship to scale and map type?

4. EXTENDED EXERCISE: Mental Maps

Objectives: Communication with maps
Things and events as patterns and processes
Accuracy is related to use

Overview

Mental maps are a way of portraying geographic relationships and features, but can only communicate in limited ways. In this exercise you will prepare a mental map of the area you live in. Later, you will find a map of the same area on the Internet and compare the two maps in terms of what they communicate and their suitability for navigation.

Instructions

On a piece of plain white paper draw your mental map of the area from memory. Don't just draw (or copy from) a street map! This should be a mental map, not a cartographic map. Take about 20 minutes to draw and annotate (for your instructor's sake) your map. Show as much detail as you can, and use a different color for the annotations if you can. Remember to focus on making the map accurate only in terms of what is important to you—the places you live, eat, work, walk, recreate, and so on. Leave off things that are not important. Include the following elements and symbols:

Bus stop △	Landmarks: prominent points of interest
⌐‾	Pathways: paths, streets, etc.
⬭ Campus	Districts: downtown, dorms, etc.
★ Park	Nodes: meeting places, centers where pathways cross
∿∿∿ Cliff	Barriers: Obstructions

Don't forget to put your name and date on the map when you are finished.

Communicating with Maps

Find a road map of the same area on the Internet. You can find maps for most areas at several websites including *mapquest.com*, *maps.yahoo.com*, and *www.multimap.com*. Print the map out if you can. What are the key differences between your map and the online map? Are both maps showing the same things? How can you explain the differences? Does it have something to do with the reason for making each map? Which map is better for communicating?

Patterns and Processes

One of the most interesting things about maps is that because the paper and drawing won't change by itself after you make it, you have to show things, such as a house or store, in the same way as you show processes—for example, the way you walk to a bus stop or drive a car to work. If you use multiple colors you can separate things and events, but remember the map doesn't show the process, it only shows an event that corresponds to the process.

Accuracy Is Also a Question of Use

Comparing the two maps, it seems to make sense that neither map is better than the other for communicating. If you want to explain to someone the place you live in, your mental map is much better in communicating the places you like, where you live and work, and what is significant in this area for you. If you just needed to explain to a visitor how to get to campus, downtown, a store, or park, than the online map is probably better suited because it focuses on giving the information needed for navigation in the area.

The potential use of a map is an important factor in determining the map's accuracy. While the online map is more accurate for general navigation, your mental map may be better for explaining to a visiting relative how to meet you at the local park or café you frequently go to.

Questions

1. What do you personally consider to be the most important features you drew on your map? Why are they important?
2. Are there blank areas on your map? If so, why? What do you guess is in these "empty" spaces?
3. How long have you lived in the area? How has this affected your mental map?
4. Do you use a car? A bicycle? How does this affect your mental map?
5. How does your mental map compare to the road map? Consider differences in detail and the use of the maps for navigation. What purposes do you think each map is better suited for?

Chapter 2

Choices in How We Make Representations

Any representation of the world is always an abstraction. It reduces complexity, simplifies, and highlights essential things, events, and relationships. An artist's painting, Hollywood movies, and a child's drawing of home are obviously abstractions. Geographic information and maps are also abstract representations, but they may be less obviously so, for a multitude of reasons. Most people believe maps are more accurate than paintings, for instance, and they wouldn't question a map's representation the way they would a painting's. One reason for this is that maps generally follow unstated rules and notions, or conventions. The established conventions for geographic information and maps explicitly and implicitly guide people in making choices and reinforcing ideas that geographic information and maps are more accurate. Geographic information and maps follow many conventions to ease understanding in order to facilitate communication. In other words, abstraction isn't "bad"—it's necessary for the sharing of knowledge and information. However, even with conventions that implicitly influence many aspects of cartography and geographic information, many map and geographic information abstractions remain difficult to understand until we have specific contextual information—for example, a electrical utility's power-line map may come with explanations, but most of us would not understand the map or the explanations.

It is certainly true that a great deal of the ease of "reading" maps comes from an individual's familiarity with conventions. Once you have understood the conventions, understanding maps is much easier and it even becomes hard to imagine how one could go without the insight they offer. An infinite number of choices are possible when creating geographic information and maps, so conventions play a key role in limiting choices to make understanding easier. You can find out about the specifics of geographic information

FIGURE 2.1. The world-at-night image is abstract. At any one time it is physically impossible to see the entire world at night. Second, the location of many lighted areas is interpolated.

Source: *http://svs.gsfc.nasa.gov/vis/a000000/a002200/a002276/index.html*

representation types in Chapter 9. How well geographic information and maps communicate greatly depends on how they apply established conventions. The world-at-night image in Figure 2.1 would be much harder to understand if the outlines of the continents were less clear: it would be difficult to see which parts of a continental landmass aren't strongly lit if the continents themselves were not shown in a shade of gray. Understanding the principles that are implicit in conventions can also help you understand maps better and prepare you for working with geographic information and maps.

This chapter familiarizes you with the underlying concepts of geographic representation and cartographic representation and cartographic communication. For now, you should keep the two types of representation separate from each other in the following sense: *geographic representation* deals with how people choose aspects of the world to show on a map or as geographic information; *cartographic representation* is the process and choices involved in going from a geographic representation to the symbols to communicate with readers. You need to be aware that as we become more versed in the issues, they often melt into one. Many people refer to them together as "modeling," other people will distinguish "models" from "cartography." As we learn the specifics of geographic information and cartography, holding them apart will help you to learn the underlying concepts and skills required, the role of conventions and the various meanings of models. These concepts are, of course, fundamental to understanding the conventions used for geographic information and maps. In the following chapters you can explore the various components of geographic and cartographic representation in relationship to communication in greater detail.

Geographic Representation

To illustrate the issues and concepts related to geographic representation, we will begin with an example that you may already have encountered yourself: flooding. You will come back to this example throughout this chapter, so you need some background information to aid you in making connections to cartographic representation and communication issues later. After presenting some background information, we will turn to the issues in creating a geographic representation using a database. As you will see, geographic representation is fundamental to GIS, but even without a GIS you can engage in geographic representation.

One river that frequently floods an urban area is located in northern Illinois near Chicago. The Des Plaines River is a muddy and slow river draining areas that were once prairie and now are largely developed. It flows into the Chicago River and then into the Mississippi. Children are afraid of the river because of legends about leeches and snapping turtles supposedly living there—never mind the pollution; most adults are more afraid of the river because the river floods frequently in the spring following winter snow melt and the first rains of spring. Experts say this is because of the greatly increased development near the river in northern Illinois, which has left areas that were previously fields and forests covered with impervious surfaces including asphalt, concrete, and houses. Previously, open ground had absorbed most of the rain and melting snow, only slowly releasing it into the nearby Des Plaines River. Now much of the melting snow and rain flows through culverts and pipes almost directly into the river, vastly increasing the flow of water that very quickly enters the river, increasing the volume of water in the river, and causing the river to go over its banks and flood lower lying areas, even those areas that previously hadn't seen flooding for many years.

The primary purpose for the geographic presentation is to conceptualize the necessary data to help answer the questions of when and where the river can flood. A very simple model may only consider observations of river bank elevation and water height. With observations of these characteristics, we can create geographic information that is the basis for analyzing where the river floods. To make sure that the observations can be related to one another, special attention to the measurements must be given. First, we need to consider the relationship between water height and river bank elevation. For the entire length of the river, we need to have a defined value that indicates the level of water that leads to a flood. This value isn't simply zero because the river banks become lower as the river flows away from its higher source to its mouth on the Chicago River. The water height and river banks need to be modeled in a relationship that also remains valid if there are changes along the river bank—for example, building a dike of sandbags. An ideal choice is a system for recording elevation independent of water height and riverbank elevation. Both measurements can be related to each other and will show that if the water height is higher than the river bank elevation, a flood results; additional data can also be added later. The elevation reference needs to be explicitly defined, something a regional, state, or national mapping agency or geodesy agency generally provides and keeps current.

And what about the question of where the river floods? The geographic representation could split the river into segments, the smaller the better, and for each segment record the average water height and river bank elevation. Using smaller segments is better because that will allow us to more accurately say where the flooding will occur, but this will require more data—which adds to the project's costs. Additionally, moving on to an issue for cartographic representation, segments that are too small will make it difficult to show where flooding occurs over the 250-mile length of the river and its tributaries on a small screen or paper because the small segments would be hard to visually distinguish. Before going on to consider more specifically how measurements and observations are geographically represented, please note that this example leaves out a number of important details—most importantly, the measurements and observations involved in determining the capacity of the river.

Measurement, Observations, and Relationships

We have at this point a simple model to indicate river flooding that considers the relationship between water height and river bank elevation. Each of the elevation measurements is related to an independent elevation reference. The river segments are related to an independent coordinate system, which can be used to assure that all observations are recorded at the same location. To place the measurements into a geographic representation used for geographic information we finally need to create attributes in a database that records the values for each river segment. The observations (field or calculated data) can then be stored for each segment. These values can then be compared to determine if and where the river floods for a particular water height.

The determination of flooding is based on a relationship: if the water height is greater than the river bank elevation, flooding results. This can be modeled mathematically as $w > e$. This relationship will never be stored in the geographic representation; we have to calculate it using the recorded characteristics. However, the stored geographic representations were only determined based on an understanding (however simplistic) of the relationship between water height and river bank elevation. The relationship is central for understanding flooding, but the information recorded in the geographic representation lacks this relationship at first—it must be determined and recorded as data in another step using the measurements.

In the end, measurements, observations, and relationships are all parts of geographic representations, but relationships are usually separate from

---

In-Depth When Does Data Become Information?

This question demands an answer that is straightforward but the issues at hand (representation, use, knowledge, meaning) are anything but clear. The resources identified in this chapter point to relevant literature. For simplicity's sake, you can say that *information is data that means something*. In other words, information makes sense, somehow.

Data is what is stored based on observations and measurements. Perhaps when the data is collected it means something, but when it is stored in a database and eventually used by other people who weren't involved in its collection, it may mean nothing at all. To make data "mean something," it has to be associated with a context—for example, the types of measurements, the rules used for making observations, and the possible uses of the data. With the Internet and online data clearinghouses, this has become commonplace.

Turning data into information is a process, beyond the summation of measurement and data collection issues. Meaning is more than the sum of attributes collected in the data, but arises in consideration of relationships in the data and relationships to other data and the context (broadly defined). Again, information has some meaning, but data only holds the potential of gaining meaning.

If you take geographic information to another place or context, most people will still call it information because of its meaning potential. This often gets confusing. To be specific, at least to summarize the discussion here, data can be (and often is) *called* information, but unless it has some meaning, it has reverted to the status of data.

---

measurements and observations. Some relationships can be determined using measurements and observations. Remember that geographic representations represent selected aspects of things and events, which means far more than merely "storing" the data. The cartographic representations that follow provide ways to show the relationships. After all, most maps don't only show us just things or events, they show us how things affect each other or can be related.

Types of Measurement

Measurements have to be stored as values to be information. Water height is stored as an interval value because it is related to a defined starting elevation of 0, which in this case is taken from the elevation reference. The storage of measurements is a key issue for creating geographic information and relies on the types of measurement developed by Stanley Smith Stevens.

Stevens developed his reference scheme in the 1940s. His intent was to offer a framework for psychologists and other social sciences that could take intrinsic properties into consideration. *Extrinsic properties* are those that are directly empirically measurable: width, height, depth, elevation, and the like. *Intrinsic properties* are characteristics that can be observed, but must be associated with other properties—for example, color, age, form, quality. Extrinsic properties can be established directly from an object, but intrinsic properties must be indirectly measured, inferred, or interpreted.

Because of the nature of intrinsic properties, Stevens proposed that measurements should be distinguished according to the ability to combine them with other measurements. For example, the measurement of a person's height cannot be meaningful combined with the measurement of his or her hair color. Following Stevens, height is an interval measure and color is a nominal value.

In all, Stevens differentiates four types of measurement, which unfortunately are not exhaustive and fail to include common types of geographical information such as radial measures of angles. The four measurements and their definitions are:

Nominal	Qualitative measurements (name, type, state)
Ordinal	Quantitative measurements with a clear order, but without a defined 0 value (small, medium, large)
Interval	Quantitative measurements with a defined beginning point (temperature, height, distance)
Ratio	Quantitative measurements that provide a relationship between two properties where the 0 value indicates the absence of the relationship (particulates mg/m^3, time to cover a distance, dissolved oxygen in a liter of water)

Sinton's Framework

Applying Stevens's measurement framework to manual cartography was relatively straightforward because of an individual's (or organization's) control of the design and drawing process. The use of computers and sharing of geographic information changed this because now data collection management and output are divided between numerous individuals and organizations. During the early days of GIS, a number of people working in this area realized that the established "art and science" of the cartographer required more detailed descriptions of cartographers' and geographers' work if people were ever to successfully automate cartography, especially if maps were to be used as the basis for analysis. "Analytical cartography," as it is called, produced a number of important approaches and concepts that became crit-

	Description	Common Examples	Map Examples
Nominal	qualitative measurements	name, type, state	
Ordinal	quantitative measurements with a clear order but without a defined zero value	small, medium, large	
Interval	quantitative measurements with a defined beginning point	temperature, height, distance	
Ratio	quantitative measurements that provide a relationship between two properties where the 0 value indicates the absence of the relationship	particulates mg/m³, time to cover a distance, dissolved oxygen in a liter of water, population density	

FIGURE 2.2. Measurement types with map examples.

ical to the success of GIS. John Sinton was an important and active contributor to this group.

Sinton devised a scheme for considering space, time, and properties in three possible roles: fixed, measured, and controlled. When space is fixed, a measuring device (e.g., tide gauge, stream gauge, NO_2, etc.) measures an attribute at a set interval of time (e.g., constantly, every 10 minutes, once a day, etc.). Measuring devices often use time intervals as the control. They produce information about a single place. People usually collect geographic information about multiple places by fixing the time and measuring characteristics of space or some other attribute, and controlling the space or attribute.

The distinction between what is controlled and what is measured is important for geographic representation. If the attribute is fixed and the space is measured, the resulting representation is a *vector representation* showing the extent of the attribute; if the space is fixed and the attribute is measured, the resulting representation is a *raster representation*. Of course, there are exceptions and limitations to this approach, which is why we say it is the way that geographic information is *usually* collected. First, point data—for example, the location of the stream gauge or a measurement of soil pH, or the location of a truck—does not show extent, but it does show

FIGURE 2.3. Stream gauge showing fixed location, measured attributes (strip chart), and controlled time (minutes/hours).

location based on the measurement of space for a fixed attribute. Second, this scheme gets very complex if even everyday objects with many measurements are considered. Third, and most importantly, Sinton's scheme does not take into account relationships. As in the river flooding example before, implicit water levels can only be related by use of common elevation base values.

FIGURE 2.4. Raster and vector geographic representations and Sinton's framework.

Choices and Geographic Representation

The choices involved with creating geographic representation are wide-ranging and often elusive because conventional ways of understanding geography lump them together. Further, some of the choices have direct and obvious consequences for cartographic representation and communication, but the consequences of others are hard to pin down. For each of the choices identified here, you will find an indication of how it is relevant (highlighted in italics) to the example of river flooding we examined earlier in the chapter. These choices are discussed in more detail in Chapter 3 and the following chapters.

Data collection	To represent a thing or event, data must be collected. The collected data must offer sufficient geographic, attribute, and temporal detail for the intended purposes and uses. *How are the measurements of water height and river bank elevation made?*
Data updates	Missing roads, changed land uses, and so on can impair reliable communication with geographic information and maps. The choice to use out-of-date source materials can gravely limit reliability. *How can changes to the river banks and drainage be considered?*
Attributes	What characteristics and qualities of things and events are included and how they are recorded make certain representations and analysis possible or impossible. *Should changes to water height be recorded as new attributes or should they replace the existing attributes?*
Coordinate system	Commonly used for geographic information, maps also make use of coordinate systems, a combination of a projection, datum, and locational reference system. The coordinate system is an especially important choice for geographic information and in some areas may even be legally defined. *What is the best coordinate system for showing the river with sufficient accuracy and detail?*
Vector/raster	Will the map emphasize the areal extent of particular attributes (vector) or the presence of particular attributes (raster)? *Recorded as vector data, the areal extent of each segment is clear, but is the raster data perhaps more advantageous for observing and measuring?*

Data combinations   When data are combined with other data, as they usually are, the choices of projection, scale, and coordinate system become especially important. For example, data from one projection can usually not be reliably combined with data from another projection. *It is of the greatest importance that water height and river bank elevation measurements be recorded geographically using the same coordinate system.*

Cartographic Representation

In traditional cartographic map making, which lacked a distinct phase of working with geographic information collected by others, cartographic representation was a choice of map elements, level of generalization, and visual variables, as well as many of the geographic representation elements. The measurements and observations could be transformed manually into a dizzying array of graphical elements, some of which still remain beyond the means of GIS or remain highly difficult even with more specialized applications. In most GIS work, some or even all of the data has already been prepared, making cartographic representation often a more distinct part of the process.

For work with both geographic information and maps, the basic choices remain the same, but it's fair to say that you will have fewer chances to work directly with observations and measurements. Because of the availability of geographic information, in most cases, people are constrained by the geographic representation. Of course, there is some flexibility, but for the most part options are decisively limited by choices already made for the geographic representation. This limited flexibility and the options that go along with it are discussed in greater detail in the following chapters. The following choices serve only to orient you and are related to the river flooding example from earlier in the chapter to help you better understand the relationship between geographic and cartographic representations.

Choices for Cartographic Representation

Projection   Every map made on a flat sheet of paper or shown on a flat screen is projected. An infinite number of projections can be made. Fortunately, in most cases at least, commonly used projections have been established. *Which common projection used in northern Illinois is best suited for the presentation of river flooding? This issue needs to be considered together with the coordinate system.*

Scale	To represent locations measured in a projection system to locations on a map or coordinates used for geographic information, the locations are scaled. Scale also constrains how things, events, and geographic relationships can be shown. *Which scale is best suited for showing the river flooding? What scale best balances detail for the cartographic products required?*
Points, lines, areas	Vector data can be easily transformed between these geometric types, expanding cartographic representation possibilities. *The river, geographically represented as a line, could be converted to an area. Would this help people understand river flooding better or confuse them into thinking the river is much wider than it is?*
Cells or pixels	Raster data can be shown either as cells, which take up an area, or as a pixel, which is single dot, much like the dots making up the image on a TV screen, which combined show an area. *If the river is geographically represented as raster data, does the presentation of the river as cells or pixels help the cartographic representation?*
Symbols	The symbols for elements from the geographic representation can be varied in terms of size, shape, value, texture, orientation, and hue. *Should flooding be shown by changing the hue from blue to red as danger increases?*

Cartographic Communication

Geographic information and maps are collected and ultimately made to communicate. How this communication works is important to think about because this reflection helps improve the geographic information and maps that we make. A more detailed consideration of the underlying issues follows in other chapters. The clearest communication results from the successful consideration of the many aspects of geographic and cartographic representation in the context of the purpose and any and all relevant disciplinary or professional conventions. After all, how the content can be communicated depends on geographic representation and cartographic representation choices. If you only count the list of choices made so far in this chapter (14), this means a total of 195 choices, which is still barely scratching the surface of the infinite potential in map design.

Before looking at the practical demands of geographic information and cartographic communication, we should consider an example of the relationship between geographic representation and cartographic representation. A common example occurs between the coordinate system and the projection. In many cases, the cartographic projection is the same as the projection used for the coordinate system. In these cases, it's clear how the relationship between the geographic and the cartographic representations makes it easier for cartographic communication. Staying with one projection means no transformation between projections is required, which limits the potential for distortions to be introduced. We need to consider how when the geographic representation and the cartographic representation use different projections, the potential distortions to areas, angles, shapes, distances, and directions can lead to many small errors or quite large errors. Combining the geographic information for monitoring river flooding in northern Illinois will be much more difficult if different projects and coordinate systems are used.

In summary, geographic representation and cartographic representation should always be considered together when examining or developing geographic information or maps for cartographic communication. Although the media for geographic information and maps are vastly different and the types of communication vary greatly, the parallels are great enough that with some exceptions the basic issues for cartographic communication apply to both.

Cartographic communication, in the most general sense, relies on distortions. As Mark Monmonier (1991) writes:

> A good map tells a multitude of little white lies; it suppresses truth to help the user see what needs to be seen. But the value of a map depends on how well its generalized geometry and generalized content reflect a chosen aspect of reality. (p. 25)

Although some cartographers find this view dismissive of cartographers' labor, it highlights how distortions are necessary for cartographic communication to succeed. Perhaps we should add the important note that this also means that we need to be especially on guard when working with material prepared under a cartographic license that is ambiguous.

For cartographic communication, we should pay special attention to the following characteristics of maps:

Scale	What is the relationship between units on the map and the same units on the ground? What simplifications accompany the scale?
Projection	What kind of distortion does the projection introduce to areas, angles, shapes, distances, and directions?
Symbolization	How do symbols exaggerate or minimize features on the map? How does the cartographic communication benefit from the chosen symbols? What is the best measurement framework? Nominal, ordinal, interval, ratio?

Generalization   How have irrelevant details for the map's purpose been filtered out? How have details relevant to a map's purpose been emphasized? How have lines, points, areas, and content been handled?

Conventions

Conventions are common and important for good reasons that have been highlighted earlier. The number of choices available—and the concerns that go along with them—may seem a barrier for activities involving geographic information and maps, but due to established conventions it is actually quite easy to read and create geographic information and maps. In most cases, conventions have already dealt with issues long before you have begun to read or create a map. Pragmatically representation and communication distinguishes three types of conventions:

- Things most people anywhere in the world know
- Cultural influences and culturally influenced knowledge
- Disciplinary or professional understanding and knowledge

For example, most adults comprehend that water is symbolically represented by the color blue. Even when you look at a map with text you cannot read, you will probably be able to distinguish water areas based on their color. However, examples like this are very, very rare. Culture exerts a powerful influence on how most people understand colors and conventions. For example, the color red, which for most people in Western countries symbolizes danger (e.g., traffic lights and fire), is the color for success in Chinese culture. The greatest number of conventions, however, come from disciplinary or professional subcultures. Disciplinary or professional groups have often developed complicated formal and informal codes for representation. Sometimes the symbols become ubiquitous through use—for example, interstate highway symbols in the United States—but many remain specific to disciplines—for example, pipe line symbols used by sanitary engineers and field crews. Often effective geographic information and map "reading" and creation go hand-in-hand with an introduction to these conventions. In some countries—for example, Great Britain, South Africa, and Switzerland—children in school learn about their country's topographic map symbols and have little problem throughout the rest of their life turning to the detailed maps made in these countries to orient themselves.

Scale and Accuracy

A key component of cartographic communication is scale. Usually scale is "defined" by established conventions, rules, or possibly even laws. It has important consequences for accuracy. Scale can serve as a proxy for more complicated representations between a geographic representation and the phenomena on which it is based. Because of scale's significance for data collection and processing, indicating the scale for geographic information provides crude but effective shorthand for understanding the accuracy of geographic information. Large-scale maps cover small areas with significant detail and accuracy. Medium-scale maps cover larger areas with less detail and accuracy. Small-scale maps cover large areas with the least detail and accuracy.

The United States Geological Survey describes the positional accuracy of their topographic products in terms of the U.S. National Map Accuracy Standard (1947) that states 90% of the points tested should fall within a fixed distance (0.02 inch or 0.5 mm) of their correct position. The points tested are only the well-defined points, which leaves open the possibility that less well-defined points are far less accurate.

In terms of positional accuracy, it would be too easy to simply say that an accuracy standard means you know what you're getting. Remember that the actual location of these points could be anywhere within the areas indicated. In this sense, while the positional accuracy standard is a quantitative measure, its interpretation is often very qualitative unless exhaustive measure is made.

Scale also impacts attribute accuracy: fewer detailed measurements and observations can be made and represented at smaller scales. Smaller scales (larger areas) must generalize and combine phenomena into cartographic features that remain visible at the output or data analysis scale. Determining attribute accuracy is more complicated and generally involves comparisons of selected points on one map or in one geographic information data set with another or field checks at the actual locations. These tests are very important as they often indicate important differences between pieces of geographic information that are supposedly of the same phenomena, but in actuality are collected and/or analyzed using very different methods.

Quality and Choices

As discussed in Chapter 1, a common way to think about the reliability of a geographic and cartographic representation is in terms of quality. The highway map is "good" if we can use it to find our way easily. Simply said, good quality geographic information or maps are useful for the purpose we create

TABLE 2.1. U.S. National Map Accuracy Standard Calculations for Three Common U.S. Topographic Mapping Scales (0.02 inch or 0.5 mm)

Scale	Accuracy in ground units
1:24,000	40 feet
1:63,360	105.6 feet
1:100,000 (Metric)	416.6667 meters

or intend to use them for. By "quality," we usually mean "reliability," but quality often simply means that a map fits the intended use. For example, you could use highway maps to figure out the average size of towns in a state, but the results would be of low quality. The concept "fitness-for-use" helps people working with maps and geographic information to get a grip on the slippery concept of quality. What this means is that we need to know the intended purposes of a map before we can decide what level of quality the map has.

At this point, it's important to recognize that the choices made for geographic and cartographic representation affect quality. The example of the differences between projections and coordinate systems is merely the tip of an iceberg that following chapters will develop in much greater detail. Going back to the river flooding example and thinking about the choices between vector and raster representations, you may grasp the significant consequences of this choice for cartographic representation and the accurate communication of flood events and how this is important for any and all geographic information and maps.

Summary

The representation of things and events from the world involves choices, which are greatly influenced by conventions. The choices are endless, making conventions critical to successful communication. Representation of things and events distinguishes between geographic representation and cartographic representation. Geographic representation is the abstraction of measurements and observations to geographic information. Sinton's framework provides a useful tool for considering different ways that things and events are geographically represented as information in terms of time, space, and characteristics (attributes). The relationships among things and events in a geographical representation are critical for cartographic representations. Cartographic representation creates maps and other visual representations and takes myriad cartographic presentation issues into consideration including scale, symbols, and graphic variables. Successful communication can be considered in terms of quality, especially the "fitness-for-use" of the geographic information or map.

In-Depth The Parts of Maps

If some people had their way, every map would always include five elements that aid in understanding by whom, why, and when a map was made. However, like all recommendations these five parts are suggestions, not requirements. The five essential elements of maps are:

Legend (special note re color)
Scale
Orientation

Neatline
Title

Additional and important elements include:

Name of author
Date map published
Explanation of purpose
Projection
Data sources
Gridlines

Review Questions

1. What is the difference between geographic and cartographic representation?
2. What are the four types of measurements?
3. What do the four types of measurements leave out?
4. What distinguishes vector from raster data in Sinton's framework?
5. Why are TIN and topology not included in Sinton's framework?
6. What is the general relationship between scale and accuracy?
7. What does "cartographic communication" refer to?
8. What kinds of measurements are excluded from the established four types?
9. What is controlled in a stream gauge (using Sinton's concept)
10. How is accuracy a qualitative indicator?

Answers

1. What are the differences between geographic and cartographic representation?

   In essence, geographic representation is the selection of observations, measurements, and choices about their coding as attributes and relationships in a database. Cartographic representation is the selection of graphical elements and abstraction of geographic information to communicate for a purpose (or multiple purposes).

2. What are the four types of measurements?

   Following the psychologist Stevens, they are nominal, interval, ordinal, and ratio. Important measurements, including radial measurements, are not included in this widely used scheme.

3. How do geographic representation choices determine cartographic representation?

   How data is collected and stored as part of the geographic representation process limits the possibilities for making cartographic representations. For example, a multilane highway represented as a single line can never be used to indicate how many lanes of traffic are slowed down by heavy traffic.

4. What distinguishes vector from raster data in Sinton's framework?

Vector data measures space and controls the attribute; raster data controls space and measures the attribute. Both types of data fix time for the geographic representation.

5. Why are TIN and topology not included in Sinton's framework?

TIN and topology represent relationships. Sinton's framework involves space, time, and attributes. Relationships are at most implicit.

6. What is the general relationship between scale and accuracy?

The larger the scale (the smaller the area a map represents), the greater the accuracy can be. The smaller the scale (the larger the area a map represents), the lower the accuracy can be. The main issue here is the possible size of the printed page. More detail can be fit onto a single sheet of paper at a large scale than at a small scale. This does not apply to GI. But because of the continued use of data collection processes used originally for producing printed maps, scale continues to offer a useful shorthand for assessing accuracy.

7. What does "cartographic communication" refer to?

GI and maps ultimately communicate. How they communicate depends on the cartographic representation choices. *"Cartographic communication" refers to the ability of GI or maps to communicate for specific purposes.*

8. What do databases, in an abstract sense, contain?

Attributes and relationships used for cartographic representation and communication.

9. What is controlled in a stream gauge (using Sinton's concept)?

A stream gauge controls time, measures attribute, and fixes space.

10. How is accuracy a qualitative indicator?

Some parts of accuracy may be quantitative, but assessments of accuracy also depend on the consideration of the potential use of the GI or map, which may be only partially specified, leading to a qualitative assessment of accuracy.

Chapter Readings

Chrisman, N. R. (1997). *Exploring Geographic Information Systems*. New York: Wiley.

Gould, P. (1985). *The Geographer at Work*. London: Routledge.

Gould, P. (1989). *Becoming a Geographer*. Syracuse, NY: Syracuse University Press.

Hartshorne, R. (1939/1956). *The Nature of Geography: A Critical Survey of Current Thought in the Light of the Past*. Lancaster, PA: Association of American Geographers. [Reprinted with corrections, 1961]

Hartshorne, R. (1958). The Concept of Geography as a Science of Space, from Kant and Humboldt to Hettner. *Annals of the Association of American Geographers*, 48(2), 97–108.

Monmonier, M. (1991). *How to Lie with Maps*. Chicago: University of Chicago Press.

Monmonier, M. (1993). *Mapping It Out: Expository Cartography for the Humanities and Social Sciences*. Chicago: University of Chicago Press.

Monmonier, M. (1994). Spatial Resolution, Hazardous Waste Siting, and Freedom of Information. *Statistical Computing and Statistical Graphics Newsletter*, *5*(1), 9–11.

Monmonier, M. (1995). *Drawing the Line: Tales of Maps and Cartocontroversy*. New York: Holt.

Web Resources

Geographic representation is an important topic for research in GIScience. See *www.spatial.maine.edu/~max/UCGIS-Rep.pdf*

A prominent GIS software producer's view on geographic representation can be found at *www.esri.com/software/arcgis/concepts/gis-data.html*

The application of cartographic representation to mapping in Africa offers important insights into pragmatic issues. See *www.africover.org/carto_standard.htm*

A discussion about changes to cartographic representation helps one to think about connections to geographic representation. See *www.questia.com/PM.qst?a=o&d=5000969871*

The U.S. EPA has some interesting materials on the differences between raster and vector at *www.epa.gov/region02/gis/gisconcepts.htm*

These presentation materials provide a succinct overview of accuracy issues: *www.epa.gov/nerlesd1/gqc/courses/images/kirkland.pdf*

Part of an online training course offered by the USGS covers basic database principles in terms of spatial analysis. See *http://geology.er.usgs.gov/eespteam/GISLab/Cyprus/database_modeling.htm*

Exercises

1. Geographic Representations: Measurements, Observations, Relationships

Think of some environmental or social issues of which you are aware and consider how to represent them geographically. Take into account how you would measure the observations and relate the measurements (as geographic information) to each other. What measurement types could you use for the observations?

2. Geographic Representations: Considering choices

Starting with the questions provided in the list of choices, consider some of the potential issues in making the geographic representation you started in In-Class Exercise 1 of this chapter. Any issues are valid in this exercise. Write them down and see how they show themselves in the following chapters.

3. Quality and Choices

Based on the discussion of geographic representation, cartographic representation, and the relationships between them and to cartographic communication. Discuss on hand of the example you have worked with for the other in-class exercises, what potential impacts some of your choices can have on quality.

4. EXTENDED EXERCISE: Representations

Objective: Identify different measurements and geographic representations

Overview

A key part of creating and working with GI and maps is identifying the underlying measurements and choices in geographic representations. This can be difficult, but extremely worthwhile.

Instructions

Using maps from the library or Internet sites (e.g., *www.davidrumsey.com*), identify at least two different thematic maps. Examine the maps and identify the measurements, geographic representations, and cartographic representations.

Questions

1. What is the title, subject, and date of the map? Who created the map? Is it part of an atlas, series, or report? Where did you find the map?
2. What alternative measurements could have been made? Explain at least two measurements and the consequences for the map.
3. Is it clear how the measurements are collected? Does it say when and by whom? How is this information (or lack thereof) important for assessing the geographic and cartographic representations?

Chapter 3

GI and Cartography Issues

"All maps lie" is a statement that reflects on the necessity of relying on abstraction in order to communicate with maps. After considering geographic information and maps in terms of geographic representation, cartographic representation, accuracy, and quality, it's clear that a map's lie is disputable. Some maps overtly distort things and events and hold little similarity with the world they represent. Other maps, by simplistically applying conventions and frameworks, may greatly but unintentionally distort things to an even greater extent. Choices made in the geographic representation and cartographic representation of geographic information and maps determine accuracy and quality. Whether there is "too much" distortion comes down to a map's fitness-for-use.

This chapter examines the fundamental choices of geographic representations and cartographic representations covering the key properties: projections, scale, symbolization, and color. It lays the foundation for more detailed examinations of cartographic and geographic information principles in the following chapters. These chapters focus on these fundamental choices with the goal of introducing their application and significance for geographic information and maps.

From a Round to a Flat Surface: Projections

Representing and communicating geographic information and maps on flat surfaces (screens or paper) requires transforming three-dimensional locations from the earth to two-dimensional locations on a flat plane, which can either be a two-dimensional (2-D) coordinate system (in the case of geographic information) or a piece of paper (in the case of a map). Projections for large areas usually use a simple sphere; for smaller areas, where accuracy gains in importance, the projection uses an ellipsoid which locally corre-

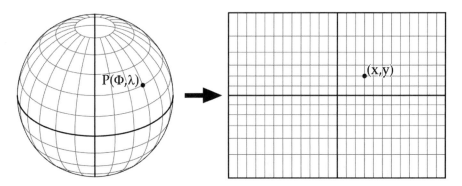

FIGURE 3.1. Most projections of locations transform three-dimensional locations into two-dimensional locations.

sponds to a *geoid*, the name for the shape with the most accurate correspondence to the actual oblate and irregular shape of the earth at a given time.

Projections have great importance for geographic information because most geographic information records the location of things and events on a two-dimensional coordinate system, called a "Cartesian coordinate system" when the *x* and *y* axes intersect at right angles (see the right side of Figure 3.1). Polar coordinates, which record location in terms of one distance and an angle from a central point, are also used. Any projection of location from the round surface to a flat plane causes some form of distortion. This has important consequences for the accuracy of geographic information or maps and what you can do with a particular projection.

Traditionally, most books on cartography start by discussing projections. Projections are one of cartography's most important contributions to science and civilization. Projections are, and have been, the foundations for almost all representations of the earth or any part of the earth. Almost all geographic information also uses projections. The ancient Greek geographer and astronomer Ptolemy invented several projections that were used by the Romans and by others for centuries afterward. Some centuries later, when European exploration and colonization commenced, because they were so important to the accurate determination of a ship's location and showing geographical relationships between mother countries and colonies, projections quickly became an important mathematical activity. You can even think of the 400 years between Mercator's publication of his global projection in 1568 and 1968 as the "golden" years of projections. Although the choice of these years is somewhat arbitrary, it roughly coincides with the period of significant European colonization and ends soon after computers made the calculations for projections a much easier task. Before moving on to the concepts of projections, you should also know that while it is possible to record the location of things and events in three-dimensional coordinate systems, they are still rather uncommon in most of geography and cartography. They are very uncommon because of their relative complexity, the widespread use of two-dimensional coordinate systems, and the cost of transforming two dimensional coordinate systems. Chapter 4 will take a look at some of these systems, including their applications.

Key Concepts of Projections

Projections convert measured locations of things and events in three dimensions to two dimensions. Projections are important but also complicated because it is impossible using geometric or more complex mathematical methods to simultaneously preserve both the shape and the two-dimensional area of any three-dimensional object found either on the spherical surface of the earth, in the earth, or near the earth, when we depict it in a two-dimensional coordinate system. Each projection is an abstraction of the earth's surface and introduces distortions that affect the accuracy of the geographic information or map. A projection starts with one of three representations of the earth's irregular surface (geoid, ellipsoid, or spheroid) and converts it directly or through intermediary transformations to a flat, or planar, transformation.

Choosing the right projection is important for controlling these distortions. Thankfully, choosing the right projection for a particular area is a task that has often been done by institutions and governments and made part of

FIGURE 3.2. World map from 1801 using a Mercator projection.

From *www.davidrumsey.com*. Reprinted by permission of David Rumsey.

FIGURE 3.3. Illustration from 1862 showing 15 projections.
From *www.davidrumsey.com*. Reprinted by permission of David Rumsey.

conventions or even laws that state what projection must be used for certain areas and activities (see Chapter 5). This is usually a good thing, but many institutions and governments require multiple projections.

Whatever you do with geographic information or maps, you need to know some projection concepts in order to understand projection distortions and their consequences. Some geographic information is stored in latitude and longitude coordinates and can be displayed or mapped on a flat screen or piece of paper, but these "unprojected geographic coordinates," as they are usually called, have tremendous amounts of distortion when shown on a flat plane.

Four fundamental concepts are crucial to know when you use geographic information and maps:

1. The earth is almost round, and always changing shape. Three models of the earth are used in making projections: sphere, ellipsoid, and geoid. A perfectly round object, or *sphere*, is defined by the mathematical relationship between the center of the object and its surface, the radius. The surface of a sphere is a constant distance from the object's center. This is the simplest model used in projections and is sufficient for geographic information and

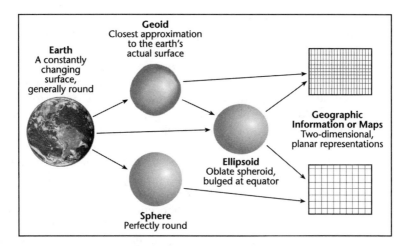

FIGURE 3.4. Abstractions of the earth used in making projections.

maps of very large areas. However, because the spinning of the earth creates a centrifugal force that causes the earth to bulge at the equator and flatten at the poles, the distance from the center of the earth to any point on the equator is greater than the distance between the center of the earth and the north or south poles. This more precise shape is known as an *ellipsoid* (but often called a *spheroid*) and comes much closer to describing the actual shape of the earth. It is accurate enough for most geographic information and maps of smaller areas. Because of different weights of material in the earth's core, differences in magnetic fields, and movements of the earth's tectonic plates, very detailed measurements of locations use a geoid for projections. A *geoid*

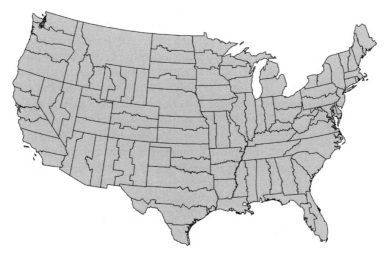

FIGURE 3.5. U.S. continental State–Plane Zones (NAD83). These zones are commonly used in the United States for state geographic information activities and are often defined by statute.

is the most accurate representation of the earth's surface. It accurately describes the location of objects to a common reference at a certain point. The difference between the sphere, ellipsoid, and geoid at any place can be as much as several hundred meters (yards). The ellipsoid and geoid models of the earth are defined and updated at irregular intervals. Should you become involved with very detailed and accurate measurements of location, you should also be aware that the geoid of the earth is constantly changing and locations recorded with an older geoid may not match a newer geoid. (See Plate 1 for geoid undulations.)

2. A projection makes compromises. Every projection either preserves one projection property or makes some compromises between projection properties. In either case, some projection properties are compromised by every projection. Because there are theoretically an unlimited number of projections, it is important to organize projections by projection properties. Which projection is used in making geographic information or a map has much to do with how geographic characteristics and relationships are preserved. The four projection properties, along with the cartographic terms in parentheses for each, are:

Angles	Preservation of the angles (including shapes) of small areas (conformal)
Areas	Preservation of the relative size of regions (equivalent or equal area)
Distance	Partial preservation of distance relationships (equidistant)
Direction	Certain lines of direction are preserved (azimuthal)

Most projections preserve area, although a large number are *compromise projections*, which means that they sometimes preserve area, but sometimes preserve shape. Usually compromise projections are used for showing the globe, but they can be used for smaller areas. All things considered, the projections that preserve area are more common because people usually need maps of smaller areas where geographic relationships and area comparisons are very important. However, the projections showing the globe are significant because they are the only way for almost all people to see and understand the world. Global projections make very significant trade-offs between projection properties. One of the most common projections used for showing the entire world, the Mercator projection, is a classic case of how a projection always trades off among projection properties. In the case of the Mercator projection, it preserves the shape and distance relationships of small areas, but only locally; it preserves lines of constant bearing; it fails to preserve area (the sizes of Greenland and Africa are greatly distorted); it partially preserves continuity, breaking Eurasia into two halves. These trade-offs mean that the Mercator projection is a good choice for representing small areas and large areas, but only for navigation.

3. Distortions will occur. Every projection, in making trade-offs between

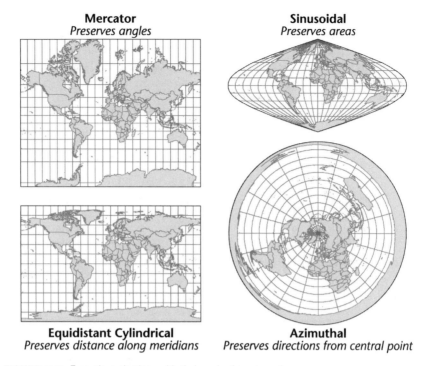

Mercator
Preserves angles

Sinusoidal
Preserves areas

Equidistant Cylindrical
Preserves distance along meridians

Azimuthal
Preserves directions from central point

FIGURE 3.6. Example projections with their projection properties.

the various projection properties, creates distortions. These distortions can be minimized by choosing a projection that corresponds as well as possible to characteristics of the area to be mapped and the known purposes and uses of the geographic information or map. Inappropriate and erroneous choice of projections can lead to significant errors and misrepresentations. Since there are no rules for choosing optimal projections, you simply have to assess each projection individually and learn through practice and discussion with other people what projection is best for a particular area, purpose, and use. In many places the projections of most geographic information and maps have already been determined. However, different people, institutions, and countries may use very different projections for the same area, requiring you to know the distortions that different projections create.

4. Geographic information from different projection should not be combined. Geographic information is particularly prone to errors resulting from the combination of data from different projections. This also applies to maps, but since it is very time-consuming to trace two maps and overlay the tracings, in practice you should be most concerned with the consequences of combining geographic information from different projections, which is perhaps one of the easiest mistakes to make with GIS. Sometimes, although you may know the geographic information is for the same place, the combined data is separated by a huge distance, possibly even many times the size of the earth. Sometimes—and this is why knowing the projection of geographic information is so important—the distances between geographic information objects can be minute, just a few inches or feet. However, because of differences in projections, what may be minute differences in one place may be vast differences elsewhere.

Assessing projection distortions and determining the best projection for an activity and area remains a complex activity that is required for working with very accurate geographic information. (See Chapter 4.)

Projected or Unprojected Geographic Information

Geographic information or maps for large areas—for example, a continent or the world—are often projected, but they can also be unprojected. If they are unprojected, the distortion is very significant because the latitude and longi-

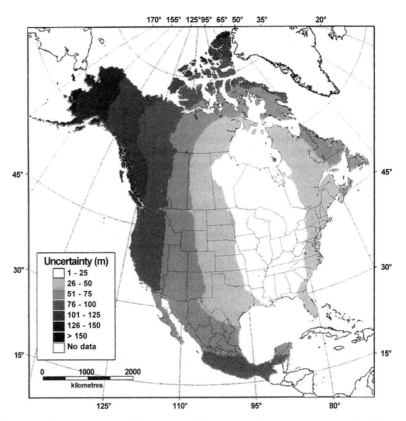

FIGURE 3.7. Positional uncertainty when coordinate system datum is unknown in North America (NAD27 or NAD83).

From Wieczorek, Guo, and Hijmans (2004). Reprinted by permission of Taylor & Francis Ltd.

tude values are converted to a two-dimensional orthogonal network of x, y values. The advantage is that unprojected geographic information can readily be transformed to other projections as needed. Smaller areas are usually projected because the projected representations better correspond to conventional maps that people have used for many years. In areas with legally established coordinate systems or with clear conventions, the choice of projection can be easy. In other areas, a few choices may be preferable depending on the orientation, size, and accepted practices for the area in question.

Projections in Practice

You should look at the distortions of the Mercator projection and the recently popularized Peters projection in Figure 3.9. The widespread use of the Mercator projection to show things and events at a global scale (which, you should note, Mercator never did) leads to very sizeable distortions, especially in areas near the poles, but also in the latitudes where most of Europe and North America are located (see Figure 3.10). These distortions led Arno Peters to promote his adaptation of older projections, the Peters's projection, which has been widely adopted even though it introduces other distortions. While the Peters's projection does not solve all projection problems, it has made people more aware of the distortions inherent in projections.

Geographic Information and Maps Are Abstractions

Finally, we should note that projection is one type of abstraction, which can be misused and even lied with. Sometimes this is obvious, but careful editing can gloss over rough spots. Geographic information and maps involve many other abstractions, which is why one of Mark Monmonnier's books on cartographic principles, uses, and abuses carries the title "How to Lie with Maps." Based on what you now know about projections, the claim that maps lie is easy enough to refute. All maps must have distortions; therefore, some would argue, what is called a "lie" is only a "distortion."

Additional Fundamental Choices

The principle choices in geographic and cartographic representation highlight key issues for the geographic information or map framework that abstracts the infinite complexity of the world. This process of abstraction is commonly called "data modeling" in GIS practice and teaching. Data modeling merges geographic representation and cartographic representation. However, when using existing geographic information, other organizations have already decided many aspects of data modeling. By keeping geographic representation and cartographic representation separate, you can develop a better understanding of the complexities of data modeling.

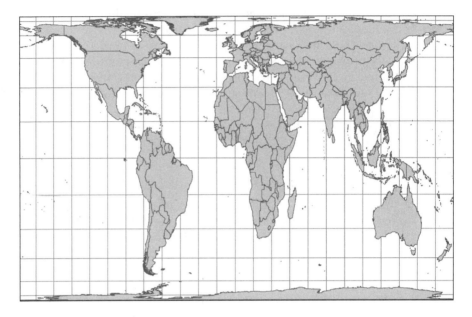

FIGURE 3.8. The Peters's projection is very similar to this Gall projection.

Things or Events

One of the most primary choices is whether to depict the phenomena of interest as things or as events. Usually, the geographic representation of things is easier. However, showing phenomena as things makes it necessary to make more complicated cartographic representations later that show the relationships among things. For example, showing a traffic jam with a symbol generally suffices to show where the traffic is stopped or slowed down. How the traffic jam develops, however, cannot be easily shown with this geographic representation. Modeled as an event, possibly at the level of individual cars and trucks, the development of the traffic jam can be represented as a dynamic process.

Patterns or Processes

The choice between patterns or processes is inseparable from the geographic representation choice of things or events. Still, while it is impossible to represent a process solely relying on things, it is possible to represent events as patterns, or to add additional geographic information to the things to show more of the process. The addition of geographic information to a geographic representation to support cartographic communication objectives—for example, how a detour for additional traffic decreased the size of the traffic jam—is a possibility for addressing some of these issues.

Abstracted or Accurate

All geographic information and maps are abstracted, but choices remain in this regard in attempting to find the appropriate balance between abstraction and accuracy. The abstraction of a cartographic representation can make the communication of important relationships easier, but diminished accuracy can make it difficult to use the resulting geographic information or map for other activities.

Few or Many Associations

Along with a balance between abstraction and accuracy, the number of associations in the geographic representation and the cartographic representation opens up some challenging issues. For communication, a map reader or geographic information user has to able to associate the final graphical product with his or her own experiences and knowledge. Sometimes, for example, in schematic maps of utility lines, the "contextual" information of a place is kept very simple and the components of the utility network are almost the only things appearing in the cartographic representation. The traditional topographic map goes in the other direction, offering a multitude of cartographic elements that can be associated in myriad ways with a person's experiences and knowledge. Most cartographic representations lie somewhere in between. Simplifying the associations can be effective in

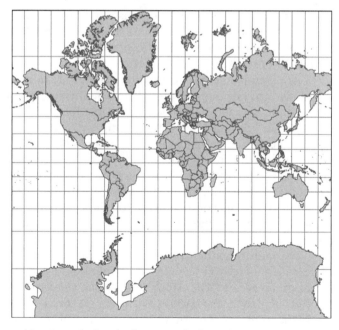

FIGURE 3.9. Mercator projection showing extremely distorted areas near the north and south poles.

focusing the representation on specifics, but runs the risk of losing critical details necessary for using the geographic information and maps. Coming back to the traffic jam example, a detailed breakdown of the traffic jam length and average speeds in individual lanes may be useful for showing relationships between entering traffic and the traffic jam, but could add too much detail for a communication goal of showing simply where the traffic jam is and how long it lasted.

Scale

The *scale*, or the relationship between a unit of distance on the screen or map to the same unit of distance on the ground, represents a critical choice. If the scale is large (showing a small area), a great amount of detail can be represented. If the scale is small (showing a large area), then less detail can be shown. (Chapter 5 presents more on scale.) This relationship of scale to detail and area constrains maps greatly and has impacts on geographic information. Although computers allow for zooming in at different scales to data, data captured at a small scale becomes very inaccurate when it is zoomed in to. What scale is chosen, whether for geographic information or maps, impacts both the geographic representation and the cartographic representation. (See Plate 2 for graphic symbols used in cartographic representation.)

Symbolization

A fundamental choice for cartographic representation is the symbolization. *Semiotics*, the study of signs, helps us to understand the meaning of symbols and how symbols take on meaning, both individually and through relationships with other symbols. Significant choices for cartographic representation involve symbols.

At a fundamental level, the choice of cartographic symbols and semiotics can be compared to a language. In the "cartographic language," a limited set of graphic variables is available for "writing" a map. Building on Bertin's earlier work on general graphic variables, several cartographers identified size, pattern, shape, color value, color hue, color saturation, texture, orientation, arrangement, and focus. The graphic variables can be combined in myriad ways, but it is clear that some variables are more associated with difference in quantities—for example, size—than others. Effective cartographic communication depends on how well the map creator matches graphic variables to the spatial dimensions of the things, events, and associations.

Through semiotics, both the cartographic creator and the reader can assess the connection between the symbols and the represented geographic things and events. The Minnesota Department of Transportation depicts traffic volumes in the Twin Cities using the three colors green, yellow, and red. The system is extremely effective in rapidly communicating traffic slowdowns. In spite of the complicated highway network, because conventional colors for traffic signals are used, people can quickly understand where traffic is slowed down or stopped. This is a meaningful use of a graphic variable to communicate a complex thing. (See Chapter 10 for more discussion about symbols and semiotics.)

Geographic representation involves choices with direct impacts on symbols and semiotics. Whether it be traffic jams, soil pH, water flow, dispersion of airborne pollutants, or household income, how the data is geographically represented plays a huge role in what symbols can be used and how the meaning and significance of the meaning can be communicated.

Color and Symbolization

The use of color in cartographic representation involves several important choices. Value, hue, and saturation are graphical characteristics of color (see Plate 3), which are most significant for pragmatic purposes (see Chapter 10 for further discussion). *Color value* refers to the different degrees of darkness or lightness of a color. High values are light and low values are dark. Color value is usually applied to distinguish ordinal data values—for example, soil pH or population density. *Color hue* is what people normally refer to as "color," which is the distinction between blue, brown, red, yellow, and so on. This distinction is a result of the reflectance of different light wavelengths by a surface. The ability to distinguish hue is commonplace among people in all cultures, but the significance of individual colors can vary widely. Color hue can be used to show nominal differences—for example, different states, types of vegetation, planning zones—but should be used very carefully for numerical values because it is difficult for people to associate a large number of hues with changes in values. *Color saturation* is the purity or intensity of a hue. Saturation is used in conjunction with value and hue to enhance reader perception of relationships and order of map features.

Organizational Structure of GIS, Software, Hardware, and Peripherals

Organizations, software, hardware, and peripherals have great influences on the possible choices for geographic representation and cartographic representation. Nowadays the various aspects of geographic representation, cartographic representation, conventions, and choices come together in the organization of GIS. A GIS always has an organizational aspect: people obtain data from other people or organizations, share the results of their analysis or mapping with other people, and coordinate the work related to GIS with other people or organizations. Additional key aspects of the organizational aspects of GIS are covered later in Chapter 12.

Key components of GIS organization are software, hardware, and peripherals. GIS *software* contains the programs, interfaces, and even procedures for processing data and making graphics and maps. *Hardware* performs the operations of the software and makes displays and printouts. The *peripherals* can be part of the core computer hardware (CPU and memory) necessary for the basic GIS software operations, but can include printers, tape drives and external hard disks, monitors, digitizers, scanners, and so on, that offer additional possibilities for working with the GIS.

Because of the complexity of the computing operations, GIS often requires supporting organizations. Even a single person working with GIS will want, or even need, help with the hardware, software, and peripherals from time to time. The size of the GIS organization often goes hand-in-hand with the size of the company or office where the GIS is being used. There certainly is such a thing as "desktop GIS," but no matter how good the marketing, the complexity of GIS necessitates good support.

Beyond people, who are often not depicted in representations of GIS organization, GIS also rely on measurements and conventions to a high degree. These aspects, often abstract and implicit, are key parts of the underlying framework for all work with geographic information and maps. Of course, without data, all the hardware, software, and peripherals will never make a GIS.

Summary

The choices made for geographic representation and cartographic representation determine accuracy and quality. Fundamental choices involve projection, scale, and symbolization. A projection is used for all accurate geographic information and maps because the coordinates of almost all geographic information is recorded in a two-dimensional coordinate system; however, the earth is a three-dimensional object. It is possible to calculate locations for geographic information or maps without using a projection, but positional accuracy is lost. Other choices involving scale and symbolization are fundamental and are often included in the data modeling that goes along with the geographic representation and cartographic representation. Scale is important, for it determines the area and detail of geographic information and maps. Symbolization leads to the meaning of symbols that should assure accuracy and quality. Color is one of the most important symbolization choices. Its use often follows established conventions. Beyond conventions, the organization of GIS including software, hardware, and peripherals can have a very strong influence on the choices that can be made.

Review Questions

1. What GIS component is frequently overlooked in descriptions of GIS?
2. What is the difference between a *spheroid* and an *ellipsoid*?
3. How is a geoid used in relationship to an ellipsoid?
4. How is scale helpful for working with GI?
5. Why would some people claim all maps lie?

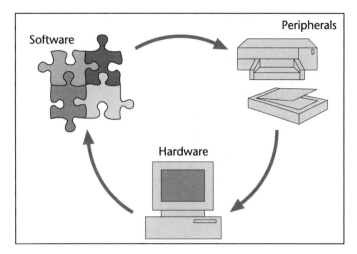

FIGURE 3.10. GIS organization as an arrangement of hardware, software, and peripherals.

    6. What are the four main choices in creating GI?
    7. What are the three components of color?
    8. What is the most common type of projection used in surveying?
    9. What projection properties does a Mercator projection preserve?
  10. What is the difference between large and small scale?

Answers

1. What GIS component is frequently overlooked in descriptions of GIS?

   People are often overlooked.

2. What is the difference between a *spheroid* and an *ellipsoid*?

   A *spheroid* is perfectly round, an *ellipsoid* is oblate.

3. How is a geoid used in relationship to an ellipsoid?

   An ellipsoid can be "fitted" to a geoid.

4. How is scale helpful for working with GI?

   Scale helps in understanding the abstractions between collection units and mapping units, as well as in understanding characteristics of the GI representation.

5. Why would some people claim all maps lie?

   All maps and GI must abstract from the actual things and events in the world, thus creating distortions.

6. What are the four main choices in creating GI?

   The four main choices are projection, scale, data, and symbols.

7. What are the three components of color?

The three components of color are hue, value, and saturation.

8. What is the most common type of projection used in surveying?

The most common type of projection used in surveying is conformal, which preserves angles.

9. What projection properties does a Mercator projection preserve?

A Mercator projection preserves direction.

10. What is the difference between large and small scale?

Small scale shows large areas, large scale shows small areas.

Chapter Readings

Bertin, J. (1983). *Semiology of Graphics: Diagrams, Networks, Maps.* Madison: University of Wisconsin Press.

Chrisman, N. R. (1999). What Does "GIS" Mean? *Transactions in GIS, 3*(2), 175–186.

Clarke, K. C. (1995). *Analytical and Computer Cartography.* Upper Saddle River, NJ: Prentice Hall.

Cotter, C. H. (1966). *The Astronomical and Mathematical Foundations of Geography.* New York: Elsevier.

Dent, B. D. (1993). *Cartography: Thematic Map Design.* Dubuque, IA: Brown.

MacEachren, A. M. (1994). *Some Truth with Maps: A Primer on Symbolization and Design.* Washington, DC: American Association of Geographers.

MacEachren, A. M. (1995). *How Maps Work: Representation, Visualization, Design.* New York: Guilford Press.

Morrison, J. L. (1994). Cartography and Organization. *ACSM Bulletin, 148,* 38–45.

Pickles, J. (2004). *A History of Spaces: Cartographic Reason, Mapping, and the Geo-Coded World.* New York: Routledge.

Robinson, A. H., & J. P. Snyder (Eds.). (1991). *Matching the Map Projection to the Need.* Bethesda, MD: American Congress on Surveying and Mapping.

Suchan, T., & C. A. Brewer. (2000). Qualitative Methods for Research on Mapmaking and Map Use. *Professional Geography, 52*(1), 145–154.

Wieczorek, J., Guo, Q., & Hijmans, R. J. (2004). The Point–Radius Method for Georeferencing Locality Descriptions and Calculating Associated Uncertainty. *International Journal of Geographical Information Science, 18*(8), 745–767.

Web Resources

A general overview of projections can be found at *http://geography.about.com/library/weekly/aa031599.htm*

For some helpful information about map projection distortion, see *www.progonos.com/furuti/MapProj/Normal/CartProp/Distort/distort.html*

For a very comprehensive set of graphics showing a great number of projections, see *www.ilstu.edu/microcam/map_projections/*

For more information on perhaps the most seminal work on projections, John

Snyder's *Map Projections: A Working Manual*, see *http://pubs.er.usgs.gov/pubs/pp/pp1395*

More mathemathical descriptions of projection are available at *http://mathworld.wolfram.com/topics/MapProjections.html*

For general information about scale, see *http://geography.about.com/cs/maps/a/mapscale.htm*

The USGS provides general information about map scale at *http://erg.usgs.gov/isb/pubs/factsheets/fs01502.html*

The importance of scale for geographic representation is handled at *http://historymatters.gmu.edu/mse/maps/question3.html*

To see how one GIS software company addresses generalization, see *www.laser-scan.com/solutions/generalisation/*

One of the best sources for practical information about the use of color in cartographic representation is *www.laser-scan.com/solutions/generalisation/*

Exercises

1. Principle Choices of Geographic Representation and Cartographic Representations

This exercise has two parts that can be broken up into two separate exercises if time doesn't allow for them to be done together.

In the first part of this exercise, take a close look at a map and describe how the map shows principle choices of geographic representation and cartographic representation. Start out by defining the purpose, the scale, and the area of the map and then consider the principle choices described on pages [xx–xx] in the book. On a sheet of paper, make six rows for the choices: things or events, patterns or processes, projected or unprojected, abstract or accurate, few or many associations, scale and areas. In one column for each row, write a few words explaining the choices as you look through the map. After you've filled in all the choices, create another column and point out the relationships to other choices. In the second part of this exercise, together with a neighbor in class, prepare a description of the choices you would make to create a map for a purpose. It could be the way to a weekend sport's event, the bus routes in your town, commuter maps, or the like. You should start out by defining the purpose, the scale, and the area of the map, and then follow the choices described on pages xx–xx in the book. On a sheet of paper, make six rows for the choices, making sure to leave a place for you to show how a sample map will look: things or events, patterns or processes, projected or unprojected, abstract or accurate, few or many associations, scale and areas. In one column for each row write a few words explaining you choice and in another column point out any relationships to other choices.

2. Detail the Fundamentals of Geography and Cartography in a Real-World GIS Application

Analyze an existing GIS application (national or global) and identify choice of projections, how GI is abstacted/derived from the data, scale choices, choice of symbols, and what GIS components are used: software, hardware, people, and organizations.

3. EXTENDED EXERCISE: Projections

Objectives: Identify types of projections
Describe the properties preseved and sacrificed by each type of projection
Relate projection to different orientations of geographic areas

Overview

Projections are crucial for geographic representation and cartographic representations. You should be able to identify different types of projections and relate them to the projection properties they maintain and to the projection properties they compromise. Finally, you should also be able to distinguish between types of projections used for different orientations of geographic areas.

Instructions

Using maps from the library or Internet sites that show different types of map projections (e.g., *www.davidrumsey.com*), identify at least four different projections. Using information from Chapter 3 on map projection properties, identify the projection properties preserved by the map projection and the map projection properties compromised by the projection. Compare the orientation of the areas shown on the different maps (north–south, east–west) and the projections used.

Questions

1. What types of projections did you identify? What are the names and subjects of the maps?
2. What projection properties do the maps maintain and what projection properties do the maps compromise? Make sure to clearly identify the names of the maps along with the names of the projection and properties.
3. Is there any correlation between the types of projections and the orientation of the geographic areas shown in the maps?

4. EXTENDED EXERCISE: Fundamental Choices

Objective: Working with maps from the library or the Internet, assess the choices made in the creation of the maps and the resulting consequences for geographic representation and cartographic representation.

Overview

The choices made in creating GI and maps determine what can be done with the GI and maps. Identify these choices as best you can and discuss alternatives for each choice.

Instructions

Using maps from the library or Internet sites that show different types of map projections (e.g., *www.davidrumsey.com*), identify two different thematic maps. Start out by describing the purpose, the scale, and the area of the map and then follow the choices described on pages xx–xx in the textbook. On a sheet of paper, make six rows for the choices, making sure to leave a place for you to show how a sample map will look: things or events, patterns or processes, projected or unprojected, abstract or accurate, few or many associations, scale and areas. In one column for each row write a few words describing the map's choice and in another column point out any relationships to other choices.

Questions

1. What are the most important choices made for each map? Describe each map and explain your reasoning.
2. What alternative choices could be made? Explain at least two choices and the consequences.
3. Did you identify any choices that lead to inaccuracies? If not, give an example of a choice for one of the maps you used for this exercise and present a choice that would lead to inaccuracies.

Part II

Principles of GI and Cartography

Chapter 4

Projections

Building on the overarching discussion of projections in Chapter 3, especially the discussion of distortion and accuracy, this chapter focuses on the principles you will need to work with projections—the transformation of spherical coordinates to planar coordinates. Chapter 4 also provides the historical background for and presents specific details of various projections. The focus of this chapter is on projections for geographic information; however, it also considers the use of projections for maps.

In fact, projections occupy one of the most essential roles in cartography for geography and geographic information. For some people, this role may arguably be perhaps the most essential, because most geographic information is "projected," even if it is never shown on a map. This has started to change as more and more geographic information is collected and stored in latitude and longitude coordinates, which are not projected. But even if all the data you need and want is available in latitude and longitude coordinates, you will probably need to project it to make the sort of map that people are familiar with.

Maps without Projections

Some people would claim that if a thing or event is shown on a map, it must be projected. In most cases this is true—and for good reasons. But there are exceptions. These exceptions are important enough to pay attention to. The first exception was already mentioned: locations stored in latitude and longitude coordinates are not projected—they are spherical coordinates. It's even possible to make a map with these coordinates, but such a map is much distorted and can even be misleading. The second exception is all the maps drawn following artistic rather than scientific concerns. Usually these maps are used for advertisements, but they can also be used to show transportation networks, to illustrate tourist destinations, and to serve other popular forms of communication. The third exception, globes, is the only non-projected way of showing things, events, and relationships without the distortion of projections.

A Brief History of Projections

The reasons for using projections go back to desires to accurately represent the spherical surface of the earth on flat maps. For geographic information and a map to be useful, the locations and relationships must be accurate. A nonprojected map using latitude and longitude coordinates, or an advertising map showing simple directions, is limited by its inaccuracy. You can use a map to the new amusement park to find your way, even if you're not from the area, but you can't use it to discover and understand the relationship of the amusement park to things and events not shown on the maps. Importantly, because of the curved surface of the earth, surveys of larger areas showing locations and sizes of things and events would be inaccurate. The Euclidean geometrical measure of the earth, which is the most common geometry—already practiced by the ancient Egyptians—cannot take curvature into account, but must be performed in a Cartesian coordinate system that has already taken the spherical shape of the earth into account. Further, using projections makes it easier to compare geographic information and maps of the same area because they provide a framework for people and organizations to systematically locate things and events.

As you can probably already imagine, it is no surprise that the first maps were based on work by geographers who were locating things. Ptolemy (c. 100–168) wrote the book *Geography* with the location of cities, coasts, and other important places of the world known to the ancient Greeks. The Romans used this book for making a map that was ultimately lost, but it was re-created in the 15th century.

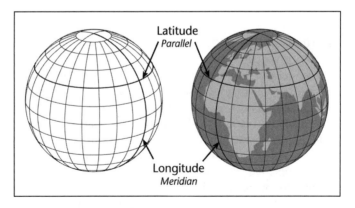

FIGURE 4.1. Latitude and longitude lines.

While maps have been used for a very long time, accurate maps of location have only been around for some 400 years, since a way of determining longitude was determined. Cartographers until then could only accurately determine the latitude of places. This means that while the equator could commonsensically be calculated as the halfway place between the north and south poles, the 0° starting measure for longitude was only agreed to internationally in the late 19th century and placed in Greenwich, England. Up until then, 0° longitudes started from different locations including Paris and the Faro Islands. Knowing where the starting measure of longitude is located is crucial for accurate navigation. Before there was widespread agreement about where the starting (0°) longitude is for all people and countries, an arbitrary starting longitude was fine so long as it was used systematically.

Roles of Projections

One of the key roles of projections has been in the production of maps for ocean navigation, which are called "charts." The development of accurate ways to determine location went hand-in-hand with the growth of European naval powers. However, because these are spherical coordinates, and mariners needed flat maps to take with them, projections became crucial. The Mercator projection is perhaps so commonplace because a straight line in this projection shows a constant compass bearing. You should remember that there are many other projections, but the Mercator projection possesses the quality that lines of a constant direction are straight lines.

Because of this character, the Mercator projection was very important for navigation on water by compass, but other modes of transportation can better use other projections. More recently, since airplanes began to fly regularly across and between continents, another type of projection was needed for their navigation. A line of a constant compass direction may be straight in the Mercator projection, but this line does not show the shortest distance.

FIGURE 4.2. Great circle path between Minneapolis, USA, and Frankfurt, Germany. The great circle distance is 4,392 miles.

Making Projections with Light

Although most projections are calculated mathematically, the underlying transformation from a three-dimensional to a two-dimensional representation of all projections can be physically constructed with the aid of few common items: a light (flashlight or lamp), a two-liter plastic bottle, a lampshade, and a piece of wax paper or flat plastic you can draw on. You will write on all of these items, so you need to be sure they are no longer needed.

To make the construction surfaces, you will need to prepare the plastic bottle by cutting off the top and bottom carefully with a scissors or knife. The lampshade and the flat wax paper or plastic are ready to be used as they are. On each of these objects you should mark a series of horizontal and vertical lines. On the lampshade and piece of wax paper or plastic, they should radiate from the center. On the lampshade, they should, if extended, meet each other at an imaginary point above the top of the lampshade; on the wax paper or plastic, they should radiate from a circle located at the center.

The construction surfaces you made correspond to the developable surfaces used in cylindrical, conic, and planar types of projections. To show how each developable surface is used, take a flashlight or light placed at the middle of the bottle or lampshade or behind the wax paper or plastic surface and shine the light source at a nearby wall or piece of paper. (It usually helps to dim the room lights when you do this.)

What you see on the wall or paper is the projected surface that corresponds to each type of projection. Try moving the light, the paper, and the construction surface to see how the changes affect each projection. These changes correspond to parameters used in the construction of map projections discussed in this chapter.

The shortest route for an airplane high above the Earth's surface is not a straight line, but a line on a sphere, called the "great circle distance."

Different projections are used for maps with different roles. The size of the area to be mapped, the desired projection properties, and the characteristics of the geographic information and map are the key determinants. The size of the area distinguishes basically between the whole world, a continent, a state or province, a region, a county or city, and still smaller units. Different projections fit different areas better or worse, depending on their use. The Mercator projection is quite inaccurate because of size distortion for world maps, but quite useful for maps of smaller areas used in navigation. A projection property refers to whether the projection represents angles, areas, or distances (from one or two points) as they are found on the surface. No projection retains both angles and areas. Most projections compromise properties, but a projection can retain one projection property—for example, the Mercator projection preserves angles. All transverse (turned 90° to be oriented north–south) Mercator projected geographic information and maps are useful for mapping north–south-orientated small areas because this projection is conformal and also preserves shapes over small areas along the line of tangency where the projection theoretically touches the earth's surface. The characteristic of the geographic information or map indicates how the projection should show geographic relationships and scales. Choosing a projection that preserves one projection property often leads to other distortions. For an individual state or province, a projection that maintains constant area to make visual comparisons of areas possible is beneficial, even if some shapes over a larger area may begin to look distorted. Indeed, larger areas are hard to show without distortion in any case; many projections commonly used for world maps compromise and distort both area and shapes. Why distortion is commonplace for projections, what are the projection properties and characteristics, and how to choose a projection is discussed later in this chapter.

Making Projections

Even if you are only going to use maps and will never work with geographic information, you need to know some important things about projections. The first is that projections use different models of the earth. Generally, projections for the entire earth can use a simple spheroid for most purposes. When dealing with maps or geographic information of the entire world, the loss of accuracy is slight compared to the resolution of the geographic information or detail of the map. Projections needed for more detailed purposes or smaller areas use an ellipsoid (also called a spheroid) that generally fits the actual shape of the earth. For very detailed purposes and the highest levels of accuracy, people use a geoid, often optimized for the shape of the earth in one particular area.

While this may seem needlessly complex, you should remember that because the earth is constantly changing shape (not only from volcanoes and earthquakes), different uses need different levels of positional accuracy. On one extreme, a map showing worldwide the most popular tourist sites needs very little accuracy; on the other extreme, an engineer's plan of a 2-km tunnel needs extremely great accuracy. Most geographic information and mapping activities need a level of accuracy somewhere in between—often cost and budget determine the accuracy.

What makes all this complicated for working with geographic information and maps is that there is no standard earth model, nor geoid model, nor spheroid model, nor ellipsoid used to represent locations on earth. The use of different models makes it paramount for geographic information users to know the model used for projecting the geographic information, which is often called a "datum" (see below for more information about datums).

The Geoid Model

The most accurate model of the earth's surface is the geoid. The earth, because of its constantly changing shape due to tectonic movements and undulations of its gravity field, can be described in the most detailed fashion through sets of measurements that are used to produce a geoid. The

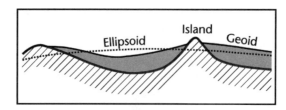

FIGURE 4.3. In this schematic, an ellipsoid and a geoid represent the earth's surface. The ellipsoid is less accurate than the geoid, but both may not properly align with actual locations.

experts who determine geoids and their constants put the geoid model into relationship with the entire planetary body or extremely detailed information about elevations in a particular area. Geodists describe a *geoid* as the equipotential surface of the earth, which means the known earth's surface under consideration of different local strengths of gravity resulting from different masses of the earth's geological makeup, fluctuations in the earth's core, and other factors. For example, the Marianna Trench in the Pacific Ocean and the large bodies of iron ore found in Sweden or Minnesota both locally affect the shape of the earth's surface because of the lessened or increased pull of gravity due to the lesser or greater mass at those locations. Basically, what geodists consider is how differences in the earth's gravity affect the shape and size of the earth. For instance, denser material in the earth's crust, such as iron, influences gravity more than lighter sedimentary rocks do. The geoid takes these (and other) differences into account. These differences are measured in millionths of the earth's normal gravity, which seems small, but the effects on the shape of the earth can be large. You also can think of the geoid as a collection of many gravity vectors, individual gravity forces, each of which is perpendicular to the pull of gravity.

Practically, the geoid was until recently only used for specialized pur-

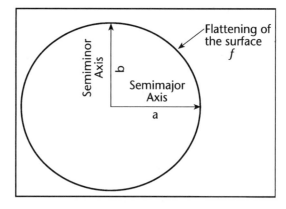

FIGURE 4.4. Reference ellipsoid showing major parameters.

poses. Geoids were almost always calculated for smaller areas because of the complexity and cost of collecting the necessary data. With the advent of satellites, however, data collection has become much easier and geoids have become more common. They are the reference standard when working with global positioning systems (see Chapter 7). The geoid provides vertical location control. Geoid positions usually refer to a reference ellipsoid for horizontal location control. Differences between the ellipsoid positions and the geoid positions are called "geoid undulations," "geoid heights," or "geoid separations." The horizontal and vertical locations of the projection surface based on an ellipsoid can be adjusted to the irregular shape of the geoid compared to the regular mathematical surface of an ellipsoid through geodetic techniques.

The Ellipsoid Model

The ellipsoid (or spheroid) is the most commonly used model for projections of geographic information and maps. It includes the noticeable distortion between the length of the earth's north–south axis and its equator, which bulges a small amount due to the centrifugal force of the earth's rotation. In the simplest mathematical form, it consists of three parameters:

- An equatorial semimajor axis a
- A polar semiminor axis b
- The flattening f

Mapmakers and geodists have produced many ellipsoids. John Snyder wrote that between 1799 and 1951 twenty-six ellipsoid determinations of the earth's size were made. Each of these ellipsoids has a history and sheds light into the science, culture, politics, and personalities involved in establishing the ellipsoid through complicated and challenging field survey coupled with exhaustive calculations. Ellipsoids were developed to satisfy individual ambition, to serve national goals, to make more accurate measurements, and so on. The surveys conducted to create ellipsoids were often ambitious expeditions into the remote areas of the world and continue to provide the material for many stories. Multiple ellipsoids were developed and refined as measurements improved and ellipsoids have often been specially defined for specific areas—for example, for U.S. counties. Working with data or maps from dif-

TABLE 4.1. Ellipsoids

Name			
Bessel (1841)	6,377,483.865 m	6,356,079.0 m	1/299.1528128
Clarke (1866)	6,378,206 m	6,356,584 m	1/294.98
Krassovsky (1940)	6,378,245 m	6,356,863.03 m	1/298.3
Australian (1960)	6,378,160 m	6,356,774.7 m	1/298.25
WGS (1984)	6,378,137 m	6,356,752.31425 m	1/298.257223563

ferent periods often involves determining if different ellipsoids were used in collecting data; data from different coordinate systems, even if in the same area, may also have different ellipsoids.

The Spheroid Model

The spheroid is the simplest model of the earth's surface, using only a single measurement to approximate the shape of the earth's surface for geographic information and maps. This measurement is the distance from the hypothetical center of the earth to the surface, or, in geometrical terms, the radius. The mean earth radius R_E is 3,959 miles (6,371.3 km). The spheroid is very inaccurate and you should only use the spheroid for scales smaller than 1:5,000,000,000. The inaccuracies of this model of the earth's surface are unapparent at these small scales. It is much easier to calculate the projections using a spherical model but using the spheroid for projecting geographic information and making maps for scales larger can lead to grave inaccuracies.

Putting the Models Together: Demythologizing the Datum

Datum is the term used to refer to the calibration of location measurements including the vertical references, horizontal references, and particular projections or versions of a projections—for example, the North American Datum 1927 or the North American Datum 1983. Datums constitute one of the most confounding aspects of working with projections for many geographic information and map users. This term simply specifies the model of the shape of the earth at a particular point in time and often for a particular area—for example, North America, Europe, or Australia. A horizontal datum is often the basis for determining an ellipsoid used in a projection for a coordinate system (see Chapter 5). A datum can be used with different projections—for example, the North American Datum 1927 is used with both the Lambert and the transverse Mercator projections. For geographic information users, datums are references to a set of parameters needed for measuring locations and the basis for projections. Because there are many parameters and the mathematics for transforming datums is highly complex, many people have been stymied by datums. But it is really, for most general purposes, quite simple: the datum refers to a reference surface for making positional measurements. While most datums in North America are described in technical guidelines or even laws, theoretically a datum can be defined by any government agency or private group as it sees fit.

Datums distinguish between horizontal and vertical references and local and geocentric datums. A datum should (but might not) contain both horizontal and vertical references. Horizontal references are used to measure the location of positions on the earth and vertical datums are used to measure the elevation of a position. You can think of a vertical datum as the base level used in recording elevations or the mean height of tides. All elevations using the vertical datum are related to this zero elevation. Local datums, in fact, are used for areas up to the size of continents—for example, the North American Datum of 1927, which made a location on Meades Ranch in Kansas the starting point of the triangulation that measured the earth's undulations and put them into relationship with the Clarke 1866 ellipsoid. Geocentric datums—for example, the World Geodetic System Datum of 1984—take the entire earth into consideration and lack an origin point; they don't have a defined datum point, but are calculated from a network of geodetic observations. The difference between local datums can be several hundred meters—for instance, between NAD 1927 and NAD 1983 in some areas of the United States. Conversions of measurements between the two systems can become quite complex. Fortunately, programs are widely available to transform between popular datums—for example, between NAD 1927 and NAD 1983—for most areas. A few important datums in North America and globally are listed in Table 4.2.

Types of Projection and Their Characteristics

Theoretically, the number of possible projections is unlimited; practically, the number is limited only by the creativity of mathematicians and geodesists and the needs of organizations to coordinate their creation, maintenance, and use of geographic information. To start, you should familiarize yourself with the three basic developable surfaces, also called "projection families," used to create map projections. *Developable surfaces*, which are an actual or imaginary drawing of the projection, were used to help cartographers visualize the projection process. They are no longer used to project maps, but they are helpful in understanding projections.

Developable surfaces can be drawn, but many projections are created without them. Projections created with developable surfaces can be demonstrated using a light hung in the middle of a transparent globe or by shining a flashlight through a portion of a globe onto the developable surface. For example, a two-liter plastic bottle, cut off at both ends and marked with a constant interval of vertical lines, with a light bulb hung in the middle to project the lines on a wall, will show how a cylindrical projection projects lati-

TABLE 4.2. Selected Datums

Horizontal Datum Name	Ellipsoid	Local/geocentric	Where used
NAD 1927	Clarke 1866	Local	North America
NAD 1983	GRS 1980	Geocentric	North America
WGS 1984	GRS 1980 with additional measurements	Geocentric	World
New Zealand Geodetic Datum (NZGD) 2000	GRS 1980	Geocentric	New Zealand

Basic Shape   Developable Surface   Projection
                     and Sphere

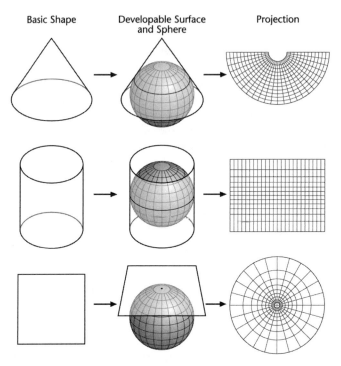

FIGURE 4.5. Basic geometric shapes (cone, cylinder, and plane) serve as developable surfaces, shown here with a reference globe. The resulting projections of latitude and longitude lines is shown in the rightmost column.

tude and longitude on a flat surface. All projections using pseudodevelopable surfaces can only be described mathematically. They cannot be created in any mechanical manner.

A key characteristic of all projections, whether developable surface or pseudodevelopable, is called "aspect" (Figure 4.6). *Aspect* refers to the orientation of the developable surface to the earth. Various conventions have come and gone in cartography over time. For future users of GIS, I think it is most pragmatic to distinguish among equatorial, transverse, oblique, and polar aspects. The differences refer either to the orientation of the projection to a region of the earth (equatorial or polar) or to the developable surface of that type of projection—for example, transverse Mercator projections are rotated 90° from the Mercator projection's usual equatorial orientation. The basic differences are best visualized in a figure showing the different aspect for each developable surface. The consequences for distortion and accuracy are discussed later in this chapter.

Some possible aspects for conical, cylindrical, and planar projection include equatorial and polar. Equatorial orientation has the projection's center positioned somewhere along the equator. Polar aspect occurs only with planar projections. All three projections may have oblique aspect (based on Jones, 1997, p. 75).

Tangent

Figure 4.7 illustrates differences in how projections "touch" the developable surface of a reference globe, another important characteristic of projections. These places of contact between the developable surface and spheroid, ellipsoid, or geoid are the most accurate for any projection and are called *standard parallels* or *standard lines*.

Projection Properties

Projections alter the four spatial relationships (angles, areas, distances, and direction) found on a three-dimensional object. Most projections only maintain one of the properties in a specific manner—for example, equidistant projections preserve distance from *one* point to all other points. Many projections, especially projections used for larger areas, compromise all these properties.

The projections that preserve angular relationships from one point are called *conformal*, but you should remember that conformal refers to the preservation of angles only, never shapes. Figure 4.8 includes a Lambert

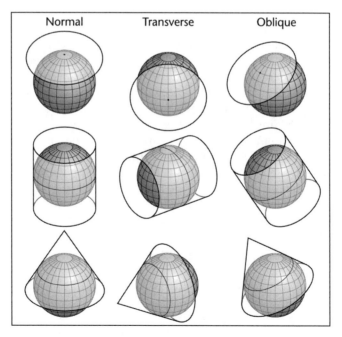

FIGURE 4.6. Some possible aspects for conical, cylindrical, and planar projections.
Data source: Jones (1997).

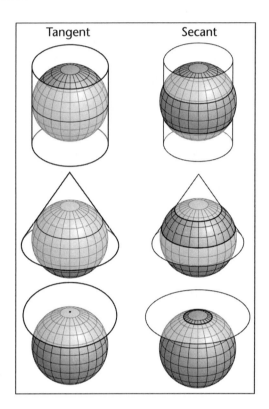

FIGURE 4.7. Examples of tangent and secant projection surfaces.

conformal conic projection, which preserves angles, but not areas. If a projection preserves areas in the projection by a constant scaling factor, it is called an *equivalent* projection. Equivalent projections preserve areas, but not shapes. The shapes of continents or countries can change in an equivalent projection, but their areas correspond to the actual areas on the earth (Figure 4.8, Sinusoidal projection). Projections that preserves distances from one or two points to other points are called *equidistant* (Figure 4.8, Stereographic projection). The projections that preserve directions are called *azimuthal*, or true direction, projections. Directions are only preserved from the center of the map in azimuthal projections.

Projections that are neither conformal nor equivalent are called *compromise* projections. They are usually developed to make more graphically pleasing maps and do this by finding a balance between areal and angular distortion (Figure 4.8, Robinson projection).

Some Common Projections, Characteristics, and Uses

With so many projections, it is possible to find a projection for every occasion. Fortunately, for most geographic information uses, the projections are already determined. The choices for maps, especially maps of large areas, are much broader. The following examples highlight a few widely used projections for each of the four projection properties.

Lambert   The Lambert conformal conic projection preserves only angles. Used for mapping continents or similar areas, it is commonly used for areas with an east–west orientation—for example, the continental United States.

Sinusoidal   The sinusoidal equal area projection preserves areas, but distorts angles and shapes. It is used for maps showing distribution patterns.

Mercator   The very common Mercator projection is a conformal projection with the very unusual quality of showing lines of constant bearing (called *loxodromes* or *rhumb lines*) as straight lines. This made the Mercator projection very

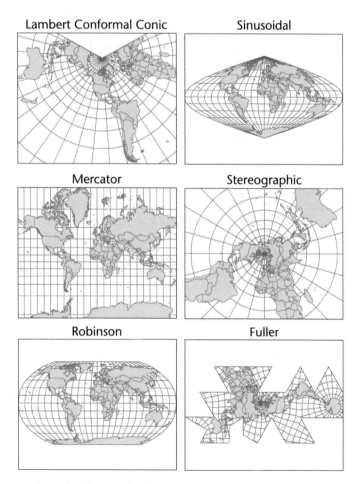

FIGURE 4.8. These six different projections show the countries of the world.

valuable for sailors, who could use one single compass heading to determine the direct route between two points.

Stereographic   The widely used stereographic projection is an azimuthal projection developed in the 2nd century B.C.E. that preserves directions; it is a further development of much older stereographic projections. It additionally has the particular quality of showing all great circle routes as straight lines; however, directions are true from only one point on the projection. It is used usually to show airplane navigation routes.

Robinson   The Robinson projection is a compromise projection that fails to preserve any projection properties. It is graphically attractive; it was adopted by *National Geographic* in 1988 and is widely used elsewhere.

Fuller   The Fuller projection was introduced in 1954 by Buckminster Fuller. It transforms spherical latitude and longitude coordinates to a 20-sided figure called the *icosahedron*.

Calculating Projections

Examining the mathematics of projections is helpful for grasping how a projection transforms locations measured in three dimensions to two-dimensional locations. You should always note that projections are never transformations between two two-dimensional coordinate systems, but between locations found on or near the surface of the three-dimensional planet earth to a two-dimensional coordinate system.

The three examples examined here are widely used. The sinusoidal projection is a pseudocylindrical projection developed in the 16th century; the Lambert conformal conic projection is widely used around the world for east-to-west-orientated areas; the Mercator projection is very common. However, the mathematics for each map projection discussed here are quite straightforward, especially since these examples are based on spheroids.

Sinusoidal Projection

The sinusoidal projection is a simple construction that shows areas correctly, but shapes are increasingly distorted away from the central meridian. Parallels of latitude are straight and longitudinal meridians appear as sine or cosine curves.

The equations for calculating the sinusoidal projection are quite simple. You only need to remember to use radians for the angle measures of longitude and latitude and to place a negative sign in front of longitude values from the western hemisphere.

Equations for calculating a sinusoidal projection

$$x = R\lambda(\cos \delta)$$
$$y = R\delta$$

Where ϕ is the latitude and λ is the longitude, R is the radius of the earth measured at the scale of map.

In-Depth **Calculating Projections with Radians**

You may need to use radians for an exercise calculating projections or for other angular measures. The sinusoidal projection, many other projections, and other measures involving angles are often calculated with radians, which is another form of angle measures: $1° = \prod/180$ radians, $360° = 2\prod$ radians. Radians indicate the length of that part of the circle cut off by the angle, and make it easy to determine distances on circular edges or round surfaces.

$$\text{radians} = (\text{degrees} \cdot \prod)/180$$

The length of part of a circle (called an arc) is determined by multiplying the number of radians by the radius. For example, the length of an arc defined by an angle of $10°$ on a circle with a 100-m radius is 0.1745.

1. Determine radian measure of angle

$$n \text{ radians} = (10° \times \prod/180)$$
$$n \text{ radians} = 0.1745$$

2. Calculate length of the arc

$$\text{arc length} = n \text{ radians} \times \text{radius}$$
$$\text{arc length} = 0.1745 \times 100 \text{ m}$$
$$\text{arc length} = 17.45$$

Some common angle measures in degrees and their equivalents in radians are listed here.

Degrees	Radians
90°	$\prod/2$
60°	$\prod/3$
45°	$\prod/4$
30°	$\prod/6$
30°	$\prod/6$

Lambert Projection

The cylindrical equal-area projection shown here is one of several projections that Lambert developed in the 18th century. It remains a widely used projection, especially in atlases showing comparisons between different countries or regions of the world. Polar areas have strongly distorted shapes, but most continents evidence only minor distortion.

Equations for calculating a Lambert cylindrical equal-area projection

$$x = R\lambda$$
$$y = R \sin \phi$$

Where ϕ is the latitude and λ is the longitude, R is the radius of the earth measured at the scale of map.

Mercator Projection

The very common Mercator projection uses only slightly more complicated equations.

Equations for a Mercator Projection (Snyder, 1993)

$$x = R\lambda$$
$$y = R \ln \tan (\pi/4 + \phi/2)$$

Where λ is the longitude (– if in the western hemisphere) for determining values of the y axis and ϕ is the latitude (+ if north, – if south of the equator). R is the radius of the earth measured at the scale of the map. The term ln refers to the natural logarithm to the base e. All angles again are measured in radians.

Distortions

Distortions arising through projections are unavoidable. They have significant consequences for accuracy, so it helps to know more about distortions in order to choose the best projection for different purposes and to be able to take distortions into account.

As a general place to start out, we can categorize distortions in terms we have already seen: the four projection properties of angles, areas, distances, and direction. Many projections distort one or two of these projection properties. Distortion of angles (including shapes) is sometimes easy to detect, especially for large areas when familiar shapes of states, continents, or even provinces are distorted; but projections of small areas may lack readily visible evidence of distortions and require the use of special graphics or statistical measures to determine the distortions. The same applies to areas. The distortions arising related to distance can be significant because, as you know now, no projection for large areas accurately shows distances for all points, but can only be accurate for a few points. Small areas are another matter, but you still should check to see what distortion a projection creates. Direction can likewise be distorted in a subtle fashion that is not visually noticeable, but is of significance should the map be used for navigation purposes.

One easily overlooked source of distortions is the difference between the datums, geoids, and ellipsoids used in creating different geographic information or maps. Even if geographic information or a map is made by the same agency or company using the same projection, a change in the datum, geoid, or ellipsoid can lead to distortions when compared with other geographic information or maps for the same area.

Describing Distortions

To describe and assess distortions, it is useful to determine the scale factor (SF) at different places on a map. By comparing scale factors with map scale at the standard point or standard lines you can assess the scale distortions using this formula:

$$\text{Scale factor} = \frac{\text{Local scale}}{\text{Principle scale}}$$

Where *local scale* is the scale calculated at a particular place and *principle scale* is the scale computed at the standard point or a standard line.

For example, the scale factor of a transverse Mercator projection with a principle scale of 1:400,000,000 calculated between 20° and 30° S will indicate how much distortion the projection introduces. First, calculate the local scale by measuring along the meridian between 20° and 30° S. This gives you the map distance, which is 3.1 cm (1.2 in.) (shown in Figure 4.9 with the letter A). You compute the ground distance between the same portion of the meridian by consulting a table showing the lengths of a degree of latitude along a meridian. At 20° a degree of latitude is 110,704.278 m long. Multiplying the 10° of latitude to 30° S would measure approximately 1,107,042.78 m or 1,107.04 km. Second, by substituting the 3.1 cm and 1,107.04 km into the map scale equation, you can calculate the local scale:

Map scale = earth distance/map distance
Map scale = 1,107.04 km/3.1 cm

TABLE 4.3. Table of Meridian Distances for Various Latitudes

Latitude (°)	Miles	Kilometers
0	68.71	110.57
10	68.73	110.61
20	68.79	110.70
30	68.88	110.85
40	68.99	111.04
50	69.12	111.23
60	69.23	111.41
70	69.32	111.56
80	69.38	111.66
90	69.40	111.69

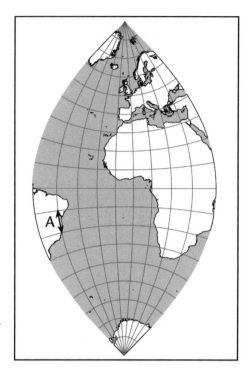

FIGURE 4.9. A transverse Mercator projection showing the location (A) for calculating the scale factor example (image scaled from original).

The units in the equation must be equal, so you first need to convert kilometers to centimeters by multiplying by 100,000.

$$1{,}107.04 \text{ km} \times 100{,}000 = 110{,}704{,}000 \text{ cm}$$

Calculate the local map scale:

Map scale = 110,704,000 cm/3.1 cm
Map scale = 1:35,710,967.74

This map scale is considerably larger than the map scale along the standard line of 1:30,000,000. You can now compute the scale factor using the scale factor equation:

$$\text{Scale factor} = \frac{35{,}710{,}967.74}{30{,}000{,}000}$$

Scale factor = 1.19

This scale factor suggests that the distances in the transverse Mercator projection increase away from the central meridian. A visual check of the projected map supports this.

Tissot Indicatrix

A visual way to examine projection distortions was developed by the mathematician Nicholas Tissot in the 19th century. The concept is simply that any small circle on a spheroid or ellipsoid, when projected to the same point on the flat map, will show the distortion created by the map projection through the projected shape and size of the circle. When the circles are plotted at various points on a map, they allow for a visual comparison of distortion (Figure 4.11). You should note that the changed shapes and sizes of the indicatrix refer to individual points and cannot be used in evaluating distortion of continents or water bodies.

The indicatrix has two characteristics that can used to evaluate distortion. The first is the two radii, semimajor (a) and semiminor (b), which are perpendicular to each other. The semimajor axis is aligned in the direction of the maximum SF and the semiminor axis is aligned in the direction of the minimum SF. The second is the angle between two lines l and m that intersect the center of the indicatrix circle, but are turned 45° in respect to the center, if there is no angular distortion. The distances of the semimajor and the semiminor axes, respectfully, indicate the scale factor distortion along each axis. The angle between two lines l and m indicates the amount of angular distortion. For example, a circle where l and m intersect at right angles indicates no distortion. If the shape of the circle is distorted into an ellipse, but the area is the same as the circle and the two lines l and m intersect at angles greater or less than 90°, there is no areal distortion, but there is angular distortion.

A map showing multiple Tissot indicatrix circles is a valuable aid to determine projection distortion. The revealed patterns of distortion help in choosing the appropriate projection for a particular area.

Combining Geographic Information from Different Projections

The large number of projections available for the same area means that great care must be taken when working with geographic information from

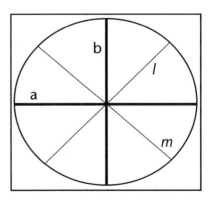

FIGURE 4.10. Tissot's indicatrix circle indicating no areal and no angular distortion.

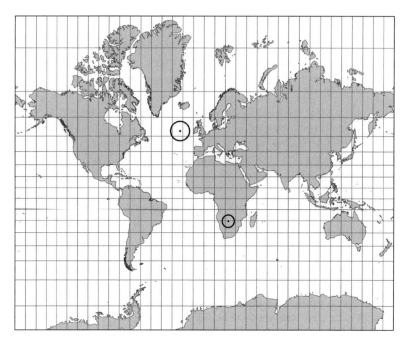

FIGURE 4.11. Two Tissot indicatrix circles shown on a Mercator projection with the standard line of the equator.

different sources. Projections for GIS provide a great deal of flexibility, but also introduce problems when working with different projections. You should note that projections used for geographic information differ from maps in an important way. When a map is made one single projection is used with a single scale for the entire map. The same thing applies for geographic information with one important difference: the coordinate system of the geographic information usually is much larger than a piece of paper used for a map. The geographic information must be scaled another time when a map is made, which can introduce some distortion. Obviously, if the geographic information is stored in the coordinates of a piece of paper, it is much harder to use it with other data, so this makes sense.

The assumption that the geographic information for the same area uses the same projection can lead to vast problems. Usually the problems when combining geographic information from different projections are so obvious that they can't be missed. Sometimes the distortions are slight and may seem inexplicable: a road from one data source is 2 m away from the property that runs along it from another data source. If care is not taken, it is possible to create great errors by combining data prepared from different projections. The same applies to coordinate systems, the topic for the next chapter, where we will look at these issues in more detail.

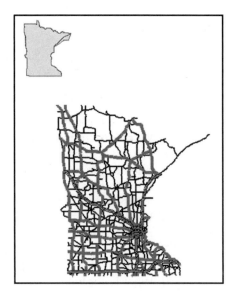

FIGURE 4.12. An example of an obvious error resulting from using data sources for the same area (Minnesota) but with different projections. Diagnosing the causes of such errors and resolving them can be very time-consuming if information about the projections is unavailable.

Summary

Projections have been the core of cartography and the basis for representing geographic information. For millennia people have developed projections to find ways to represent the three-dimensional world humans live on in two dimensions—a format much better suited for recording observations and measurements. While it is possible to make maps without a projection, unprojected geographic information or maps are greatly inaccurate and distorted. A projection can be applied to different areas and at different scales. The smaller the area, the more accurate a projection can be. How accurate the projection is depends on how the projection is constructed and what underlying model of the earth's form it uses. Basic characteristics of a projection are its orientation, tangency, and form. Projections have several properties. The most important properties are the preservation of angles (conformality) or the preservation of areas (equivalent). Only one of these two properties can be preserved in any one projection. Some projections distort both properties and are called compromise projections. The resulting distortions can be ascertained and described using a Tissot indicatrix. Because of the number of differences, it is important to assess the characteristics and properties of projections when working with geographic information, especially when combining geographic information from different sources.

Review Questions

1. Identify the type (equal-angle, equal-area, compromise) of the following projections:

   Mercator   Lambert   Mollweide   sinusoidal   azimuthal   Robinson

2. What is the difference between a secant and a tangent projection?
3. What is a transverse projection?
4. Why is a transverse Mercator projection better for north–south-oriented areas and states (e.g., Illinois) than a Lambert conformal conic projection?
5. What are the three important characteristics of projections?
6. Why is most GI projected to a two-dimensional, Cartesian coordinate system?
7. Why should you never combine GI from different projections?
8. How can positional distortion be measured?
9. What is the difference between a geoid and a spheroid?
10. Why are Mercator and Peters projections technically satisfactory? Why do people consider the Mercator projection to be a bad projection?

Answers

1. Identify the type (equal-angle, equal-area, compromise) of the following projections:

   Mercator   Lambert   Mollweide   sinusoidal   azimuthal   Robinson
   (Equal-shape) (Equal-area) (Equal-area) (Equal-area) (Equal-distance) (Compromise)

2. What is the difference between a secant and a tangent projection?

   A secant projection surface "touches" the earth's surface in two places; a tangent projection "touches" only at one.

3. What is a transverse projection?

   A transverse projection is a cylindrical projection, which is normally orientated east–west, rotated 90 degrees to a north–south orientation.

4. Why is a transverse Mercator projection better for north–south-oriented areas and states (e.g., Illinois) than a Lambert conformal conic projection?

   The conic projection works best for areas with an east–west orientation; its line(s) of tangency run east–west. The transverse Mercator projection's line of tangency runs north–south, providing a more accurate positional reference than the Lambert conformal conic projection of the same area.

5. What are the three important characteristics of projections?

   Equal-shape: preservation of shapes; equal-area: preservation of areas; equal-distance: preservation of distances

6. Why is most GI projected to a two-dimensional, Cartesian coordinate system?

Several reasons need to be considered. Much GI comes from maps with such coordinate systems. Most GI is used to make planar maps. Most GIS are designed to store two-dimensional coordinate locations.

7. Why should you never combine GI from different projections?

GI from different projections for the same area will be in different coordinate systems that do not align properly.

8. How can positional distortion be measured?

For small-scale maps, Tissot's indicatrix provides a good graphical indicator. Large-scale maps, showing small areas, require the use of statistical measures.

9. What is the difference between a geoid and a spheroid?

A geoid is a more accurate representation of the earth's surface, accounting for local variations. A spheroid is a perfectly round form that fails to account for local variations and the oblateness of the earth resulting from its spin.

10. Why are Mercator and Peters projections technically satisfactory? Why do people consider the Mercator projection to be a bad projection?

The Mercator projection is well suited for compass navigation at sea. The Peters projection is a compromise that offers a different way of representing the world. The overuse and ill-suited use of the Mercator projection to show regions of the world has led to the Mercator acquiring a bad reputation.

Chapter Readings

Jones, C. (1997). *Geographical Information Systems and Computer Cartography*. Upper Saddle River, NJ: Prentice Hall.

For a fascinating, if wide-reaching, biography and study of a person who was instrumental in determining the elliptical shape of the earth, see:

Terrall, M. (2002). *The Man Who Flattened the Earth: Maupertuis and the Sciences in the Enlightenment*. Chicago: University of Chicago Press.

For information about the basic mathematical principles of cartography, see:

Cotter, C. H. (1966). *The Astronomical and Mathematical Foundations of Geography*. New York: Elsevier.

For a history of projections, see:

Montgomery, S. (1996). Naming the Heavens: A Brief History of Earthly Projections. *Science as Culture*, 5(25), 546–587.

For a very thorough history of projections, see:

Snyder, J. P. (1993). *Flattening the Earth: Two Thousand Years of Map Projections*. Chicago: University of Chicago Press.

Web Resources

This Canadian National Atlas site offers a sound foundation on the use of projections: *http://atlas.gc.ca/site/english/learningresources/carto_corner/map_projections.html*

This JavaScript tool supports the interactive creation of global map projections: *http://atlas.gc.ca/site/english/learningresources/carto_corner/map_projections.html*

More specific information about projection parameters and accuracy is provided by government agencies, for example, the California Department of Fish and Game: *www.dfg.ca.gov/itbweb/gis/Downloads/projections/DFG_Projection_and_Datum_Guidelines.pdf*

A good resource for fundamentals of geodesy is provided by the U.S. National Geospatial-Intelligence Agency, *Geodesy for the Layman*, available online at *http://earth-info.nga.mil/GandG/publications/geolay/TR80003A.htmlinfo.nga.mil/GandG/publications/geolay/TR80003A.html*

See the Natural Resources Canada CartoCorner for a general introduction: *http://atlas.gc.ca/site/english/learning_resources/carto/index.html*

For information about homemade map projections, see *http://octopus.gma.org/surfing/imaging/mapproj.html*

Exercises

1. Projections for Different Needs

If you collect maps from magazines and newspapers for a few weeks, you will have a pretty sizeable collection of different kinds of maps and different kinds of projections.

Come up with a list of different uses of maps and the projections used for each use.

Think about how projections can preserve the shape of things on the earth, their size, or the distance from a point or along a line or must compromise between these three projection properties.

Knowing what you do now about the different qualities of projections, what do you think about newspaper maps that do not indicate the projection? Are they common? What kind of errors do you think can arise?

If possible, you also can explore the collection of maps and atlases in a nearby library.

2. Questions for Map Projections

1. Is the map whole or broken up?

2. What shape does the projection make the map?

3. How are features (continents and islands) arranged?

4. Are gridlines curved or straight?

5. Do parallels and meridians cross at right angles?

3. EXTENDED EXERCISE: Sinusoidal Projection

Overview

In this exercise you will calculate values for a sinusoidal projection that you produce.

Concepts

The location of a point (x, y) in a sinusoidal equal area projection is calculated for this exercise in two steps. First, the longitude value is transformed to east–west values (x) by multiplying the longitude value times the radius and times the cosine of the latitude. Multiplying the longitude values by a cosine of latitude creates the gradually increasing distortion of areas further away from the equator. The north–south values (y) of the projection are calculated through a linear relationship between the radius and the latitude. Second, you will scale the calculated x and y values to fit a map on a piece of paper by determining a scale ratio that transforms the radius of the sphere (6,371 km).

Exercise Steps and Questions

Preparation

In this exercise you will be calculating a projection of a graticule. You will have to do the calculations and show that you have done them, but you can work with other people to check your answers and determine the process. Before the calculating part of this exercise, let's look at the fundamental problems of projecting a spherical object on a plane.

Part 1. Angle Measures: Degrees and Radians

In Part 2 of this exercise, you will need to make the calculations in radians. Radians are one of three ways to measure angles. They are mainly used for engineering and science. We won't spend much time getting into the mathematics of angular measures. For this exercise, you only need to understand the relationship between degree and radian measures of angles.

If you know an angle measure in degrees, you can easily convert it to radians, another measure for angles used in engineering and scientific calculations:

$$radians = (degrees \cdot ?)/180$$

For example, 180 degrees equals 3.14 radians; 90 degrees equals 1.57 radians; 45 degrees equals 0.785 radians. As the examples show, radians express angular measures in relationship to the radius.

Part 2. Construct a Sinusoidal Projection of a Graticule

STEP 1: CALCULATE THE PROJECTION

Use the table below for recording the results of your calculations. The rows indicating latitude are on the left and the columns indicating longitude are on the top. You will be calculating the sinusoidal projection for latitudes 0°, 30°, 60°, and 90°, and for longitudes 0°, 30°, 60°, 90°, 120°, 150°, and 180°. Your results will be in kilometers, or, for an idealized projection surface, about 10,000 km in length and height.

The equations you will use are:

$$x = radius \cdot longitude \cdot cosine\ (latitude)$$
$$y = radius \cdot latitude$$

Where: radius = 6,371 km.

Remember: convert all angle measures from degrees to radians by multiplying by pi and dividing by 180 degrees. For example, 30° corresponds to pi/6 *using this equation*

$$radians = (degrees \cdot \prod)/180$$

Table of projected values (Step 1)

Latitude	Longitude						
	0°	30°	60°	90°	120°	150°	180°
0°	0,0						
30°							
60°							
90°							

STEP 2: SCALE THE X, Y VALUES AND THEN GRAPH THEM

The x, y values calculated in Step 1 are in kilometers; therefore they are certainly too large to fit on a piece of paper. As with creating any other map, the values need to converted to map units by determining a ratio that fits the x, y values on the sheet of paper you use (8.5 × 11 inches, or approximately 22 × 33 cm). Scale can be determined by putting the ground values and map values in the same units, here cm, and calculating the ratio between the shortest ground value distance and the longest map value distance.

Determine this value and fill it in here:

Scale factor: _____

With the scale factor, convert your original projected values to map units. Use the table below for those calculations.

Table of Projected Values (Step 2)

Latitude	Longitude						
	0°	30°	60°	90°	120°	150°	180°
0°	0,0						
30°							
60°							
90°							

Using a ruler, graph each coordinate pair on the x and y axis on a separate piece of paper. The graph should look like the northeastern quadrant of a sinusoidal projection. When this is completed, label the axis with tick marks that indicate the corresponding degree value from 0° to 90° latitude and 0° to 180° longitude. This is a map projected to a sinusoidal projection.

Questions

1. The sinusoidal projection is an example of an equal area projection. What are the major differences between this type of projection and conformal projections?
2. Why do the x values lack two-dimensional scaling at 0° longitude in the sinusoidal projection?
3. What are the major differences between the Mercator and sinusoidal projections? How big is a pole in each projection?
4. Minneapolis/St. Paul is located at approximately –93° longitude, 45° latitude. What is the x, y coordinates in the sinusoidal projection?

Chapter 5

Locational and Coordinate Systems

Applying Projections

Projections make it possible to create maps and two-dimensional geographic information using locational and coordinate systems to create common reference systems. Reference systems are used by national governments, state and provincial governments, local governments, the military, nonprofits, and businesses around the world. They make it possible for many people to work with geographic information and maps without following the specifics of projections. Their widespread use has made them crucial references for many activities. Common reference systems simplify the recording of the location of things and events and make it possible to combine information from different sources and to verify any distortions of positional measurements.

A commonly used locational or coordinate system helps greatly to minimize distortion. Accordingly, and just as with projections, working with geographic information or maps can require knowing which locational or coordinate system is used. Different locational or coordinate systems can record the location of things and events from the same area at different places. Even if they seem to overlap when drawn together, varying degrees of differences can lead to subtle or significant errors that can impact analysis.

Locational systems are different from coordinate systems. Although the terms are often used interchangeably, it is important to recognize a key difference. A locational system *can* be referenced to a projection; a coordinate system *must* be referenced to a projection and the reference model of the earth's shape and size. Usually this is known as a *datum*. Locational systems, generally with orthogonal coordinates, are only valid for particular data or possibly even with one map and may not have any connection to other locational and coordinate systems.

In this chapter you will read about the creation, history, and use of locational and coordinate systems. You will also find out about how to transform geographic information between different coordinate systems. A more detailed discussion of public administrative uses and issues is found in Chapter 12.

Locational Systems

Locational systems use a locally defined coordinate system or grid to indicate locations. These systems may even be useful for relatively large areas, but quickly run into accuracy problems due to the failure to consider the curved surface of the earth. Their usefulness is also greatly limited by the degree of adoption. That is, a locational system for a map of city parks, state fairs, downtowns, campgrounds, and so on work fine on that map, but if other people create another more popular locational system for the city, it may run into disuse and disregard.

To understand the significance of locational systems, we can begin with the Roman centuration across many areas of Europe, which is still geographically significant today. The practices of the Roman centuration, like those of

FIGURE 5.1. Archaeological finds show that the Roman survey in Great Britain corresponds to current landscape features in some areas.

From *www.sys.uea.ac.uk/Research/researchareas/JWMP/venta5.html.* Reprinted by permission of John Peterson.

FIGURE 5.2. Hierarchical subdivision of the Roman centuration.

Data source: Dilke (1985).

Egyptian surveyors, offer fascinating insights into the historical roots and centrality of locational and coordinate systems. The technical details of the centuration system also highlight concepts that societies today still rely on. The similarity between the Roman foot, at 29.57 cm, and the modern American foot, at 30.48 cm, or just about 0.9 cm different, points to the persistence of common measurements.

Roman administrators actively surveyed conquered and politically associated areas undergoing integration into the empire. The survey created new rectangular subdivisions of land that could be more easily administered and awarded to army veterans as compensation for their years of service. Evidence of centuration can still be found in areas of modern Italy, France, Tunisia, Spain, and Great Britain.

Centuration usually involved the creation of a local location system based on two orthogonal meridians. One meridian ran north–south, the other ran east–west. Based on this initial grid, the area was further subdivided into smaller and smaller units of land. Because the meridians were local and not tied to a projection and datum each centuration was a locational system.

The Public Land Survey

The Public Land Survey (PLS), also known as the Public Land Survey System (PLSS), is similar in concept to the Roman centuration. It is used in most areas of the United States to survey land for recording ownership using a grid-like system (a similar system, the Dominion Land Survey, is used in large parts of Canada). It has become very influential on the landscape of the United States and has had many impacts related to governance and administration, which are examined in Chapter 12.

The PLS was created through the Land Ordinance of 1785 and the Northwest Ordinance of 1787, following the initiative of Thomas Jefferson and the support of other surveyors. The rationale was the sweeping need to equitably provide access to land in the United States and help the government pay debts through the sale of land. After the Revolutionary War the U.S. government took on responsibility for all areas west of the original 13 states. The survey systems used prior to this time revealed themselves to have many problems that still persist. For example, the amount of land grants claimed in Georgia in 1796 was more than three times greater than the actual amount of land in the state.

All these western lands were considered to be the "public domain," except beds of navigable bodies of water, national installations such as military reservations and national parks, and areas such as land grants that had already passed to private ownership prior to subdivision by the government. This included land awarded to private individuals by the governments of France, Mexico, and Spain. Part of the original intention was the efficient allocation of land to soldiers who had fought for the United States, but the PLS was also seen to be a way to help pay off debts from the war and to cover future expenses. The original public domain included the land ceded to the federal government by the thirteen original states, supplemented with

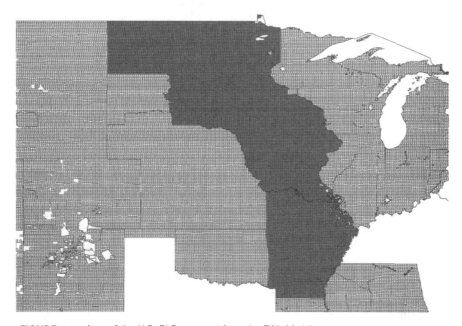

FIGURE 5.3. Area of the U.S. PLS surveyed from the Fifth Meridian.

acquisitions from Native American Indians and foreign powers. It encompasses major portions of the land area of 30 southern and western states. Almost 1.5 million acres were surveyed into the PLS system of townships, ranges, and sections.

The PLS is a hierarchical land subdivision system that makes it possible to locate land. The hierarchy begins with two of the 34 meridians, which are distinct and unrelated to other PLS meridians. The 34 principal meridians run north–south. Each is named, and allows the identification of different surveys. The meridians meet baselines, as they are called, which run east–west and are perpendicular to a principal meridian. The first subdivision of theoretically 6 × 6 mile units is organized by *townships*, which indicate the location north or south of a baseline, and *ranges*, which indicate the location east or west of a principal meridian. Each 6 × 6 mile unit is called a township and is further theoretically divided into 36 equal 1 × 1 mile sections. Each section has a theoretical area of 640 acres and can be further divided into aliquot parts including half sections, quarter sections, and quarter-quarter sections. (See Figure 5.4.) The location of all land in this system consists of

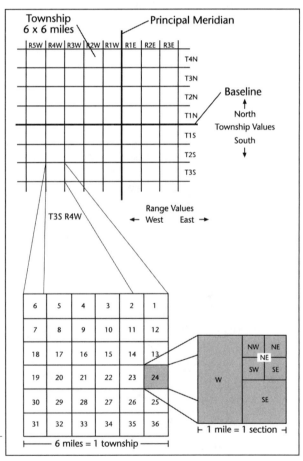

FIGURE 5.4. Nested hierarchical organization of the U.S. PLS.

FIGURE 5.5. PLS in western Washington State highlighting unusual 1/2 townships and ranges. These peculiar townships are the results of difficulties surveying in the alpine Olympic Mountain Range. Other distortions are also plainly visible in this portion of western Washington State.

the state name, the name of the principal meridian, township and range designations with cardinal direction, and the section number.

In some areas, due to survey difficulties, errors, or even fraudulent surveys, the townships and sections may vary considerably from the theoretical system of 6 × 6 mile townships and 36 equal 1 × 1 mile sections. (See Figure 5.4.) Fraudulent surveys, surveys that were indicated as correct, but in actuality contained some discrepancy, were often detected before becoming legally binding, but in some case frauds slipped through and became legally valid. In some cases, surveys may have been conducted in difficult terrain, beyond

Public Highway and Fundamental Survey Line (Section, Above; V.M.D. Unit, Below) Coincident, c. 1875.

Relationship of Public Highways and Fundamental Survey Lines in the Example Areas, c. 1875.

FIGURE 5.6. Different lengths of roads mean different lengths of bridges between systematic and nonsystematic surveyed areas.

From Thrower (1966). Reprinted by permission of Norman Thrower.

Aerial photograph of an area in northwestern Ohio subdivided in the manner of the United States Land Survey System. The black x's indicate a quarter section (1/2 x 1/2 mile).

Aerial photograph of an area in the Virginia military district of Ohio subdivided in an unsystematic manner.

FIGURE 5.7. Different location systems used in surveying have environmental consequences. From Thrower (1966). Reprinted by permission of Norman Thrower.

the capabilities of the surveying equipment or the surveyors. Errors may have been made in spite of thorough cross-checking of survey measurements. Some people hired to conduct surveys sought short-cuts to simplify the work, and produced surveys that were complete on paper, but may have distorted things on the ground. Regardless of the limitations, errors, or frauds, the markers placed by the "original" surveys are considered the ultimate authority for all later land subdivisions and locations, even if some quite complex problems required later resolution.

The PLS's consequences go beyond the creation of a system for subdividing land and the development of the ability to systematically locate land in most of the United States. The "original" survey marked the land for future development. If you have ever flown across the midwestern or western United States and looked out the window, you probably noticed a landscape that looks like a grid stretching out to the horizon. Traveling in a car on many roads in this area, you probably noticed that the road goes straight for a long time, with only minor deviations, and intersections with roads are mostly at right angles. Both are the consequences of the PLS. But there are more significant consequences, which are environmentally and economically significant, as Norman Thrower discussed. First, the PLS subdivision of land does not follow existing natural features, which usually help guide the use of land. Intensive farming practices in PLS areas can more easily have detrimental effects than the same practices in areas surveyed using metes-and-bounds surveys. The roads in the PLS may be easier to drive on, but the costs of maintaining bridges may be higher because they have to be longer. In Thrower's study, he established that 60% of all bridges in an area surveyed using PLS were longer than 20 ft (6 m), versus only 20% in areas that were surveyed unsystematically.

Local Coordinate Systems

A system like the PLS is registered to a meridian and a baseline that have known coordinate values. Further, any PLS locations can be associated with other coordinate systems. While it is possible to determine coordinate locations in the PLS, the system functions without any reference to coordinates associated with the earth's size or shape.

Local systems are further removed from relationships with the earth's size or surface. Although the Roman centuration relied on meridians, which were surveyed based on astronomical observations and measurements, the survey of PLS meridians makes the relationship with the earth's size and surface and with other meridians secondary, meaning for surveying and legal purposes that a portion of the PLS is essentially a local location system. Smaller local coordinate systems are very commonplace because they are very handy for quickly aiding people to use and orientate themselves with maps. However, they are of no use for recording the location of things and events when they should be used with other locational and coordinate systems.

RECTANGULAR COORDINATE SYSTEMS

Rectangular coordinate systems are different from locational systems in that they are associated with a particular model of the earth's size and shape (geoid or ellipsoid). A datum is usually also associated with a particular projection. However, this broader use of the term has become more commonplace.

The PLS system, when associated with a rectangular coordinate system,

FIGURE 5.8. Figure showing an arbitrary local locational system for Minnesota highways and towns.

takes on characteristics of both locational and coordinate systems, although surveyed locations never replace the legally binding locational system. In other words, for mortgage lenders, title insurances, and banks, the locational system description is what is important—the associated coordinates have little significance.

METES-AND-BOUNDS SYSTEMS

Metes-and-bounds systems, used historically most often for local maps that record the location of parcels, were often connected to local meridians and parallels. They can be associated with a projection and made into rectangular coordinate systems with great ease. Most metes-and-bounds systems are nowadays connected to a rectangular coordinate system. In the United States, metes-and-bounds is the legally recognized system for recording parcels in the areas of the original 13 colonies and Texas, plus the areas of a few other states. In most areas of the world, metes-and-bounds systems are the more common system for recording not only the extents of land parcels, but also of legally registering land ownership, rights, and responsibilities. The metes-and-bounds system can either start with recognized origin points and then survey the boundaries of parcel boundaries based on distance and angle relationships, or just survey boundaries based on existing surveys. The former is the preferred approach, as it avoids many inaccuracies that lead to significant land conflicts.

Metes refer to the distances and angles. *Bounds* refer to the corners and points that define the outline of the surveyed area. A metes-and-bounds description is a narrative that describes the clockwise or counterclockwise path around the perimeter. A simple example of a metes-and-bounds description can read like this:

> Beginning from the southwest corner of section, thence north 1,320 feet; thence east 1,735 feet to the true point of beginning thence east 500 feet, more or less to State Road 35 right-of-way, thence northwesterly along said right-of-way.

A more detailed metes-and-bounds description can also describe the vicinity of the surveyed area, exempted areas, and additional rights to areas described in the survey. The following is an example of a more detailed description, of the diagram in Figure 5.9. Once facing due South or North, the numbers in the parentheses are the offset angle (in degrees, minutes, and seconds West or East) using bearings of the surveying equipment.

> Beginning at the concrete monument, thence S (83 deg 58' 06" W) for 211.19 ft along the North right-of-way of the highway; thence N (18 deg 40' 10" E) for 150.00 ft along the East line of Brown; thence S (72 deg 21' 10" E) for 170.00 ft along the South line of Smith; thence South for 68.00 ft along the West line of Jones, back to the point of beginning.

FIGURE 5.9. Simplified example of a metes-and-bounds description.

From *www.premierdata.com/literature/Intro Land Information.pdf.* Adapted by permission of Premier Data.

State–Plane Coordinate System

The State–Plane Coordinate System (SPCS) is a system for specifying positions of geodetic stations and measuring location using plane rectangular coordinates. This is a local coordinate system for each state which is legally defined in federal and state law. This coordinate system divides all fifty states of the United States, Puerto Rico, and the U.S. Virgin Islands into over 120 numbered sections, referred to as "zones." Larger states generally have multiple SPCS zones—for example, Minnesota has three and California has six. Each zone has an assigned code number that defines the projection parameters for the region. SPCS uses three projections, depending on the orientation of the zone and the state. The Lambert conformal conic projection is used for areas with an east–west orientation. Areas with a north–south orientation use a transverse Mercator projection. The area of the Alaskan panhandle uses an oblique Mercator projection.

The SPCS uses two datums, the North American Datum of 1927 and the North American Datum of 1983 (NAD 1927 and NAD 1983, respectively) based on different models of the earth's shape and size. NAD 1927 uses Clarke's 1866 spheroid (equatorial radius 6,378,206, flattening 1/294.98); NAD 1983 uses GRS 1980 (equatorial radius 6,378,137, flattening 1/298.26). The differences between the datums' geoids are significant and lead to sizeable differences (up to several hundred meters) between locations recorded using SPCS NAD 1927 and SPCS NAD 1983 (see Chapter 3). Since then, numerous regional modifications have also been made. These changes necessitate great care when working with geographic information from the

TABLE 5.1. The Parameters for the Three SPCS Zones in Minnesota Using the 1927 North American Datum*

SPCS Zone	Semimajor Axis (m)	Semiminor Axis (m)	Southern Standard Parallel	Northern Standard Parallel	Longitude of Origin	Latitude of Grid Origin
North	6378206.4	6356583.8	47 02 00	48 38 00	–93 06 00	46 30 00
Central	6378206.4	6356583.8	45 37 00	47 03 00	–94 15 00	45 00 00
South	6378206.4	6356583.8	43 47 00	45 13 00	–94 00 00	43 00 00

*See *http://rocky.dot.state.mn.us/geod/projections.htm#27MSP* for further information.

United States. Geographic information collected with the later NADCON or HARN improvements will differ from NAD 1983 data.

The U.S. National Grid

The U.S. National Grid (USNG) was standardized in 2001 in response to needs for a single coordinate system for the entire United States, especially for location-based services for mobile phones, GPS, and other navigation devices. The USNG can be extended to include coordinates for locations anywhere in the world. It is not intended to replace the SPCS, nor other coordinate systems, but it provides a coordinate system with national scope. It is a hierarchical system, using the grid system specified in the Military Grid Reference System (MGRS). The coordinates are also identical with UTM coordinates in areas of the United States. Coordinates can be specified in two precisions. For example, the location of the Washington Monument in Washington, DC, is:

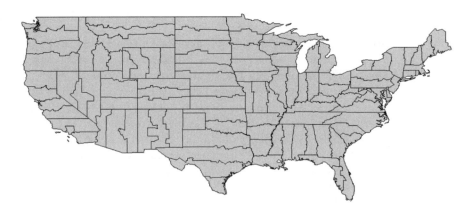

FIGURE 5.10. State–Plane Coordinate System (SPCS) zones using the North American Datum 1927 (NAD 1927).

General reference: 18SUJ23480647–precision 10 m

Special application: 18SUJ2348316806479498–precision 1 mm

The U.S. geographic area is divided into 6-degree longitudinal zones designated by a number and 8-degree latitudinal bands designated by a letter. Each area receives a unique alphanumeric Grid Zone Designator (GZD)–for example, 18S. Each GZD 6 × 8 degree area is divided into a systematic scheme of 100,000-m squares where a two-letter pair identifies each square–for example, UJ. A point position within the 100,000-m square shall be given by the UTM grid coordinates in terms of its easting (E) and northing (N). The number of digits specified the precision:

18SUJ20	Locates a point with a precision of 10 km
18SUJ2306	Locates a point with a precision of 1 km
18SUJ234064	Locates a point with a precision of 100 m
18SUJ23480647	Locates a point with a precision of 10 m
18SUJ2348306479	Locates a point with a precision of 1 m

Universal Transverse Mercator

The Universal Transverse Mercator (UTM) grid was developed in the 1940s by the U.S. Army Corps of Engineers. In this coordinate system, the world is divided into 60 north–south zones, each covering a strip 6° wide in longitude. These zones are numbered consecutively beginning with Zone 1, between 180° and 174° west longitude, and progressing eastward to Zone 60, between 174° and 180° east longitude. The conterminous 48 United States are covered by 10 zones, from Zone 10 on the West Coast through Zone 19 in New England. In each zone, coordinates are measured north and east in meters. The northing values are measured continuously from zero at the equator, in a northerly direction. To avoid negative numbers for locations south of the equator, the equator has an arbitrary false northing value of 10,000,000 meters. A central meridian through the middle of each 6° zone is assigned an easting value of 500,000 meters. Grid values to the west of this central meridian are less than 500,000; to the east, more than 500,000.

Other National Grids

Most countries in the world have one or more national grids, analogous to the U.S. system. In the world there are thousands of these systems. The following examples are exemplary for different approaches to organizing coordinates. The United Kingdom, for example, has a hierarchical system that begins with a grid of 100 100-km cells, identified by two letters. Each 100-km cell is further divided into 100 10-km grid cells. A 10-km grid cell is further divided into 100 1-km grid cells. Germany uses a system similar to UTM, but

FIGURE 5.11. Example of a location in the USNG coordinate system.

From *www.fgdc.gov/standards/documents/standards/xy_proj/fgdc_std_011_2001_usng.pdf*

bases it on 3°-wide stripes at the 6, 9, 12, and 15 meridians. The zones are numbered two through five, or the meridian longitude divided by three. A false easting of 500,000 meters is calculated for east–west coordinates in each stripe, and north–south coordinates are the distance to the equator. North–south coordinate values have seven digits and east–west coordinate values have six digits, precluding switching the coordinates. Australia has developed new national grids directly as coordinate systems at frequent intervals, reflecting both frequent tectonic movement (up to 7 cm/year) and improvements in geoid measurements. The Geocentric Datum of Australia coordinate system is the most recent and is based on the ellipsoid measure-

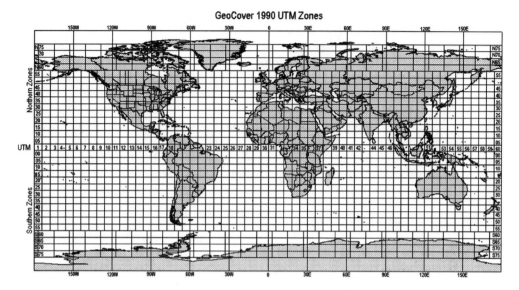

FIGURE 5.12. UTM zones.
From *https://zulu.ssc.nasa.gov/mrsid/docs/gc1990-utm_zones_on_worldmap.gif*

ments from GRS 1980 and coordinates from the International Earth Rotation Service.

Smaller countries generally use only one projection and geoid for the entire country. People who live here (and use only local maps) may never even have to learn about projections and be concerned with how to combine data from different projections.

Polar Coordinate Systems

Polar coordinate systems are necessary in areas around the poles, but can be used for specialized applications in other areas as well. A two-dimensional polar coordinate system records locations based on an angle measurement (azimuth) from the central point, the pole, of the coordinate system and a distance to that point. A three-dimensional coordinate system records location with two angle measurements and the distance to the measured point from the center. One angle measurement records the horizontal angle on the XY plane, the other records the angle on the Z plane.

Spherical Coordinate Systems

A basic spherical coordinate system records the location of things and events using three values: x, y, and z. x stands for the east–west coordinate value, y for the north–south coordinate value, and z for the elevation in relationship to a reference height. It is similar to a three-dimensional polar coordinate system, except that the origin point lies at the center of the coordinate space—for example, the theoretical center of a sphere.

A two-dimensional plane of x, y coordinates that correspond to latitude and longitude coordinates is often used by GIS to represent the entire world at once, but it introduces such grave distortions that it should only be used for browsing.

Global grids follow a different approach to creating a global grid, usually based on hexagons or octahedrons, to subdivide a sphere hierarchically into smaller and smaller triangular facets. These coordinate systems are still rather uncommon—they are mainly used for satellite tracking and studying global processes—although the advantages of these systems are significant.

Scales and Transformations

Any map you will ever see has a scale. It may be only implicit, as in a graphic artist's rendering of a summer festival site, or a city's advertising map, but more often you'll find explicit scales. An important question for the use and creation of geographic information and maps is: What is the appropriate scale? A scale too small, that shows a large area, will require that small specific things and events be removed, whereas a large scale may lead to important contextual information being left out. To work well with scale it is critical to familiarize yourself with different ways of representing the relationship between a distance unit of geographic information on a map and the corresponding distance unit on the ground.

Scale is shown for geographic information and maps in three ways:

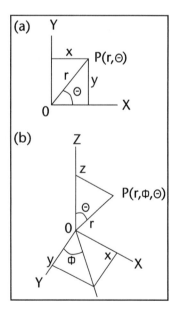

FIGURE 5.13. Two-dimensional polar coordinate systems.

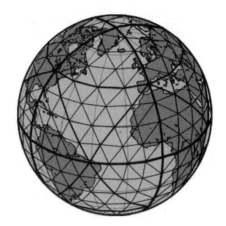

FIGURE 5.14. Global tessellation.

From *www.spatial-effects.com*. Reprinted by permission of Geoff Dutton.

- Representative fraction
- Scale bar
- Statement

The three types are equivalents, but have different representations. A *representative fraction* provides a ratio between the same units of measure on a page and on the ground. A *scale bar* graphically represents distinct distances at the scale of the geographic information or map. A *statement* describes the scale in words. The most important thing for representing scale is that the measurement units on the page (or for the geographic information) and on the ground must be kept the same. For example, the representative fraction scale 1:24,000 indicates that 1 inch on the map corresponds to 24,000 inches on the ground. Divide by 12 (the number of inches in a foot) and you'll have the basis for the statement of scale: "1 inch equals 2,000 feet." Using metric units, the calculations are even easier: the representative fraction scale 1:25,000 indicates that 1 cm on the map corresponds to

FIGURE 5.15. Representative scale and scale bars from a USGS map.

25,000 cm on the ground. Divide by 100,000 (the number of cm in a km) to determine the statement of scale "1 cm equals 250 m or a quarter km."

Scale Transformations

GI, whether collected in the field, collected from existing geographic information, or digitized from existing maps, can be readily transformed to other scales. The scaling of geographic information may be helpful for many reasons. Most often, scale transformations allow the association of any arbitrary coordinates from known places—for example, building corners or street intersections—to be associated with coordinates of the same places in other coordinate systems. In this way, locations of things and events drawn on a piece of paper can be transformed into geographic information using a coordinate system.

Scale transformations allow for an infinite number of alterations to shapes and changes. They can change all axes by the same factor, each axis by different factors, locally vary the transformation values, or use logarithmic factors. These different types of scale transformations are necessary to support the different type of changes to coordinates required when working with geographic information from different sources.

Several things need to be considered for working with scale transformations. First, it is important to remember to keep using the same units throughout the transformation. Geographic information locations stored in

TABLE 5.2. Representative Scale and Equivalent Ground Distances

Scale	Ground Distance
Standard (inches)	
1:2,400	200 ft
1:20,000	1,667 ft
1:24,000	2,000 ft
1:62,500	approximately 1 mile
1:63,360	5,280 feet (exactly 1 mile)
1:125,000	approximately 2 miles
1:800,000	approximately 8 miles
Metric (centimeters)	
1:1,000	10 m
1:2,500	25 m
1:10,000	100 m
1:25,000	250 m
1:50,000	500 m
1:100,000	1,000 m (1 km)
1:250,000	4,000 m
1:500,000	50,000 m (5 km)
1:1,000,000	100,000 m (10 km)
1:2,000,000	200,000 m (20 km)

metric units should be kept in metric units. If a transformation is made between metric and standard units, be sure that all geographic information was converted using the same constants. The transformations can also alter geographic representations and cartographic representations, leading to geographic information that is not only inaccurate but also incorrect. A common example is scaling small-scale maps to match large-scale maps of the same area. Because the small-scale maps lack accuracy in comparison to a large scale map, differences between the two maps can be the results of changes made during the generalization process—for example, when a road is displaced to fit the railroad track symbol in next to a bend in a river.

A Sample Scale Transformation

The simplest type of sale transformation is an affine transformation. Even an affine transformation makes it possible to scale, rotate, skew, and translate geographic information coordinates.

Affine transformations use two equations for the x and y coordinates of two-dimensional geographic information.

$$x' = Ax + By + C$$
$$y' = Dx + Ey + F$$

The values x and y stand for the coordinates of the input geographic information; x' and y' stand for the coordinate values of the transformed geographic information. A, B, C, D, E, and F are the six geometric parameters for transforming the geographic information coordinate values. Some GIS require the entry of these parameters; others will calculate them for you based on common reference points in the input geographic information and in the output geographic information. A linear transformation simply multi-

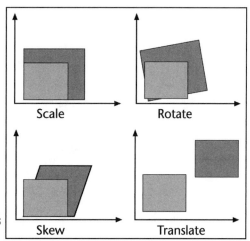

FIGURE 5.16. Affine scale transformation operations (generalized).

plies the coordinate values by the scale factor to obtain the scaled geographic information.

Summary

This chapter turns to location systems and coordinate systems, both of which often involve a projection, but may be developed without any reference to the earth's size or shape. Location systems are more likely to be locally developed ways for describing location using a grid of letters and numbers. Coordinates without a reference to the earth's size and shape are a type of location system. Coordinate systems may have a reference to the earth's size and shape through a projection, normally described as a datum. Location systems are important because they are very common and can be used to coordinate activities. Land subdivision is an important activity involving both location systems and coordinate systems. It establishes the divisions of land used in determining ownership. The U.S. Public Land Survey is possibly the most widely used systematic survey for subdividing land. More common in the rest of the world are unsystematic surveys that use metes-and-bounds approaches to recording the boundaries of land parcels. Because of their importance, law often specifies coordinate systems; usually these are called national grids. In the United States, the State–Plane Coordinate System is the best example. The newer U.S. National Grid is another example of a

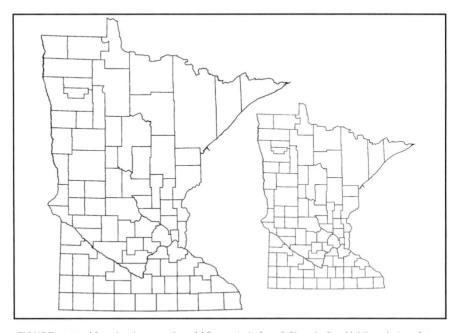

FIGURE 5.17. Map showing counties of Minnesota before (left) and after (right) scale transformation.

Accuracy of Georeferencing

According to a recent article in the *International Journal of Geographic Information Science*, six steps are required for determining georeferences to places lacking precise coordinate references—for example, 6 miles NW of Timmons, NV. Knowing these six steps is important because many places with descriptive data lack accurate locational references; however, coordinates must be used for storing the descriptions as geographic information. The point–radius method summarized here provides consistent and accurate interpretations of locality descriptions and identifies potential sources of uncertainty.

Step 1: Classify the locality description. The quality of the description should be assessed and classified. Only somewhat accurately described localities should be georeferenced.

Step 2: Determine coordinates. Coordinates can be retrieved from gazetteers, geographic name databases, maps, or from local descriptions with coordinates—for example, field notes with GPS coordinates. The numerical precision of coordinates should be preserved during processing to minimize the propagation of error. Next, *identify named places and determine their extents.* Every named place has an extent. This should be determined in the same manner as the coordinates of the locality. Most named places have a geographic center (courthouse, church) which should be used as the origin of circle defining the extent. Then, *determine offsets.* Many localities are located by their relationship to another place—for example, 6 miles NW of Timmons. The direction from the place can usually be inferred, considering environmental constraints and additional information in the description. Supplementary sources are helpful.

Step 3: Calculate uncertainties. In this article, Wieczorek et al, consider six sources of uncertainty:

1. Extent of the locality. The maximum extent of two places in the locality is the maximum uncertainty.
2. Unknown datum. The differences can be as large as 500 m between NAD27 and NAD83. Theoretically the difference could be as large as 3,552 m.
3. Imprecision in distance measurements. Treat the decimal portion of distance measurements as a fraction and multiply the distance measurement by this fraction. Multiples of powers of 10 should be multiplied by 0.5 to that power of 10.
4. Imprecision in direction measurements. Translate cardinal directions to their degree equivalents, using half of that degree equivalent as the uncertainty.
5. Imprecision in coordinate measurements. Consider latitude and longitude error.

$$uncertainty = lat\_error^2 + long\_error^2$$

6. Map scale. Take the error of a map to be 1mm. For example, the uncertainty for a map of scale 1:500,000 is 500 m.

Step 4: Calculate combined uncertainties. The uncertainties without directional imprecision and combined distance and direction uncertainties should be calculated following map accuracy guidelines for the maps used. Distance uncertainties should take directional imprecision into account.

(cont.)

Step 5: Calculate overall error. Assuming a linear relationship between individual errors and total error, use a root-mean-square equation and apply the law of error propagation to determine the maximum potential error.

Step 6: Document the georeferencing process. Documentation of the process and considerations used in determining the georeferencing are important for people working with the locality information later.

Based on: Wieczorek, J., Q. Guo, et al. (2004). The Point–Radius Method for Georeferencing Locality Descriptions and Calculating Associated Uncertainty. *International Journal of Geographical Information Science, 18*(8), 745–767.

legally mandated coordinate system. Universal Transversal Mercator (UTM) is widely used around the world, and thus is of great importance. All location systems and coordinate systems use a scale to reduce the size of measurements on the ground to map size or a comparable size in geographic information. Geographic information from one scale can be easily converted to another scale; it is much more labor-intensive to scale different maps.

Review Questions

1. What are common applications for spherical coordinate systems?
2. What is the main practical importance of coordinate systems?
3. What is the main difference between coordinate and locational systems?
4. What is the transformation from *x*, *y* to *x′*, *y′* called when all scale factors are the same?
5. What is the difference between rectangular and polar coordinates?
6. For what purpose was Roman centuration devised?
7. What is the similarity between metes-and-bounds and the PLS in the United States?

TABLE 5.3. Common U.S. Surveying Measurements

1 link = 0.66 feet or 7.92 inches
1 pole or 1 rod = 16.5 feet or 25 links
1 chain = 100 links, 4 rods, or 66 feet
80 chains = 1 mile, 320 rods, 1,760 yards, or 5,280 feet
1 acre = 10 sq. chains, 160 sq. rods, 4,840 sq. yards, or 43,560 sq. feet
1 square mile = 1 section of land or 640 acres
Township = 36 sq. miles (36 mile-sq. sections)

These survey measurements are historical and archaic, but because of their legal nature these historical surveys are still valid. Current surveys generally use standard or metric measurements.

8. What is the State–Plane Coordinate System?
9. How are locational and coordinate systems used for public administration?
10. Why are 3-D coordinate systems still uncommon?

Answers

1. What are common applications for spherical coordinate systems?

   Spherical coordinate systems are commonly used for satellite tracking and global models.

2. What is the main practical importance of coordinate systems?

   Coordinate systems provide for the common recording of positional locations against which distortions can be measured.

3. What is the main difference between coordinate and locational systems?

   Coordinate systems are mathematically defined based on a model of the earth's surface and shape; location systems can be mathematically defined, but are usually created without relating them to a model of the earth's surface and shape.

4. What is the transformation from x, y to x', y' called when all scale factors are the same?

   A constant scale transformation is called a linear transformation.

5. What is the difference between rectangular and polar coordinates?

   Rectangular coordinates are orthogonal, that is, the x, y origin is defined by a right angle.

6. For what purpose was Roman centuration devised?

   Colonizing and developing conquered areas.

7. What is the similarity between metes-and-bounds and the PLS in the United States?

   In the areas they respectively dominate, they are legally accepted means of recording land ownership.

8. What is the State–Plane Coordinate system?

   The State–Plane coordinate System was established during the 1930s in the United States to specify coordinate systems for each area. Many states have legally adopted the SPCS.

9. How are locational and coordinate systems used for public administration?

   Locational and coordinate systems are used for recording locations, helping government/private coordination, and providing a structure for future activities.

10. Why are 3-D coordinate systems still uncommon?

    The complex mathematics and lack of a common reference standard along with the abundance of 2-D maps and GI hinder the widespread use of 3-D coordinate systems.

Chapter Readings

Caravello, G. U., & P. Michieletto. (1999). Cultural Landscape: Trace Yesterday, Presence Today, Perspective Tomorrow for "Roman Centuriation" in Rural Venetian Territory. *Human Ecology Review, 6*(2), 45–50.

Dilke, O. A. W. (1985). *Greek and Roman Maps.* London: Eastern Press.

Ferrar, M. J., & A. Richardson. (2003). *The Roman Survey of Britain.* Oxford: Hedges.

Goodchild, M. F., & J. Proctor. (1997). Scale in a Digital Geographic World. *Geographical and Environmental Modelling, 1*(1), 5–23.

Linklater, A. (2002). *Measuring America: How an Untamed Wilderness Shaped the United States and Fulfilled the Promise of Democracy.* New York: Walker & Company.

Thrower, N. J. W. (1966). *Original Survey and Land Subdivision.* Chicago: Rand McNally.

Web Resources

NOAA's National Geodetic Survey maintains key resources for locational and coordinate systems at *www.ngs.noaa.gov/*

The documentation of the U.S. State–Plane Coordinate System prepared by James Stem is available from NOAA at *www.ngs.noaa.gov/PUBS_LIB/ManualNOSNGS5.pdf*

Roman Centuration is described in some detail by John Peterson at *www.sys.uea.ac.uk/Research/researchareas/JWMP/AgrimensoresMapConv.pdf*

The National Atlas provides an introduction to the PLSS at *http://nationalatlas.gov/articles/boundaries/a_plss.html*

Detailed instructions for the Public Land Survey are contained at *www.blm.gov/cadastral/Manual/73man/id1.htm*

A very thorough and well-developed introduction to the history and legalities of land surveys and information in the United States is available at *www.premierdata.com/literature/Intro%20Land%20Information.pdf*

A good description of the metes-and-bounds survey system from Tennessee is available at *www.tngenweb.org/tnland/metes-b.htm*

Detailed description and parameters for the State–Plane Coordinate System are available at *www.ngs.noaa.gov/PUBS_LIB/ManualNOSNGS5.pdf*

Resources on U.S. datum measurements and conversions are available at *www.ngs.noaa.gov/PC_PROD/pc_prod.shtml*

The U.S. Forest Service and Bureau of Land Management maintain a website for information for specific PLS questions at *www.geocommunicator.gov*

The Information and Service System for European Coordinate Reference Systems (CRS) has a website at *http://crs.bkg.bund.de/crs-eu/*

Exercises

1. Use PLS Coordinates

On a topographic map of the place you're from or one you're familiar with in the United States, determine the location of your home, school, and other important local feature using the Public Land Survey coordinates. Townships should be given on the east and west edges of the map, ranges on the north and south.

2. EXTENDED EXERCISE: Locational and Coordinate Systems

Overview

In this exercise you will interpret the impact of land subdivision systems and learn how to read coordinates from United States Geological Survey (USGS) topographic maps. This exercise also introduces you to basic topographic mapping concepts and how to recognize them.

Concepts

Topographic maps are the most general-purpose maps in circulation. They represent a multitude of features and relationships that you can "read" by looking at and studying a map. Among other things, they are useful for studying general land use development.

Exercise Steps and Questions

Step 1. Find a USGS 7.5 Minute Topographic Quadrangle

In this step you should use an index of topographic quads for a state in the United States to find a map of your home or a place you are familiar with. The feature you look for can also be a particular place, mountaintop, radio antenna, or structure—for example, a lighthouse. It should be small enough to be located distinctly: a small building or pond is OK, but not a large lake or structure. Most of all it, *it should be someplace you are familiar with*. If you can't find the 7.5 quad you need, first check to see if an older quad is available for the area. If not, choose a feature somewhere else.

Answer these questions before continuing:

1. What is the name of the place you chose?

2. What is the type of feature?

3. Why did you choose it?

4. What is the latitude and longitude of the southeast corner of the map?

5. What is the distance, in kilometers and miles, from east to west across the map?

Kilometers _____

Miles _____

6. What is range of elevation in this area?

   Highest _____

   Lowest _____

   Average _____

7. Does this map show elevation in feet or meters?

8. Do you see *any* consequences of land subdivision—for example, in the orientation of roads?

9. What is the name of the map you choose?

Step 2. Locate Your Feature

10. Find the feature you described in question 2 and locate it using the following coordinate and land subdivision systems:

    Latitude _____   Longitude _____

    UTM Northing _____   UTM Easting _____

    Township _____   Range _____ Section _____

    You should use a straightedge for these measurements and interpolate the distance.

11. What is the datum of the UTM coordinates?

Chapter 6

Databases, Cartography, and Geographic Information

Geographic representation and cartographic representation abstract observations and measurements about things and events in the world. These abstractions are only useful if they are saved and stored in formats that enable access to them. Databases provide the most common computerized means to save and store data, and they also use encoding and organization to access, manage, and analyze the data. Databases follow various principles to create a structure that is extremely flexible for the needs of geographic representation and cartographic representation; however the structure is one that can be daunting in its complexity. Databases are fundamental to systematizing representations of the world. At the same time, the use of databases also opens new possibilities for creating many different representations. Geographic information created years earlier now can be accessed and combined with other geographic information if the databases are accessible.

This chapter focuses on providing a concise overview of relational database technologies as they are used for GIS maps and geographic information. It introduces the basic principles of relational databases, forms of storage, and applications. The objective is the presentation of a solid overview of database technology in terms of issues for geographic representation and cartographic representation, including data modeling.

What Is a Database?

A *database* is a collection of data stored in a structured format using a computer. A database can be thought of as a table, but the distinction is that the table is just one way (of many) to represent the database.

The first databases were flat–file databases: computer files of text with one record on each line, usually encoded in the ASCII format. Each entry (e.g., a person's name and address), was separated by a special mark and commas or tabs separated the characteristics (e.g., name, street, house number, city, post code, state, and country). Finding a particular person required searching through the entire data file one entry after another. A flat–file database can also be represented as a list, a table, or a spreadsheet. In other types of databases the data is stored as records and fields that correspond to entries and characteristics in the flat–file database. Records and fields have become the accepted terms when working with databases. The term "tuple" is used to represent a single data item in a table. A "field" refers to the division of the data into separate parts of each data item. An "attribute" is the particular entry in a field (e.g., Main Street in the field "Street"). A single database record including all attributes is called an "entity."

The relational database sets itself apart from flat–file database through the way it stores data and the possibilities for relating data. A relational database stores data in separate files, usually called "tables," which can be related to other tables in the database by common fields. A relational database may consist of 100s or even 1000s of tables. Every table in a relational database has a key field that allows each record to be uniquely identified. This key field is usually indexed to speed up operations. This key field is especially important for geographic information because of the large amounts of data that easily come together. In a relational database, attributes can stand in different relationships to attributes in other tables. A one-to-one relationship relates a single record in one table with a single record in another table. A one-to-many relationship relates a single record in one table with multiple records in another table; a many-to-one relationship does the opposite. A many-to-many relationship relates many records from one table to many records from another table. This last relationship is rarely desirable because the meaningful relationships between the records cannot be differentiated from spurious and erroneous relationships. A one-to-many or many-to-one relationship may be called for in a variety of situations (e.g., the cities of one

Name	Example	Organization
Flat file	John Andrew, 123 Main St., Anywhere, IL, 60022 USA | Sally Andrew, 123 Main St., Anywhere, IL, 60022 USA |	
Relational	1 John Andrew, 123 Main St., Anywhere, IL, 60022 USA 2 Sally Andrew, 123 Main St., Anywhere, IL, 60022 USA	

FIGURE 6.1. Simplified representation showing the most common types of databases for geographic information and maps.

state or the states of one country). These usually reflect a hierarchy or a grouping of attributes and their corresponding records.

The relational database has several advantages for geographic representation and cartographic representation. First, the conceptual model of the database is distinct from the physical model (how the database is stored and managed on computer hardware). Second, separate tables help maintain the integrity of the potential meaning of database elements. Most relational databases now use structured query language (SQL) for constructing queries involving tables of a single database, or with tables in other databases, even on other computers. Third, the clarity of the relationships aids people using the database with previous experiences of the database. Reliable processing is critical for queries of geographic information and online maps. Fourth, it is possible to define multiple views of the same data in different database tables (e.g., listing entries by street address or alphabetically by name).

While the relational database is the most common type of database and possibly the only type of database you will ever work with, two other types of databases may be significant. The first of these is a hierarchical database. This database is organized by defining a hierarchy into which all data is stored (e.g., country, state/province, county, municipality). This type of database was frequently used for business transactions, but is being replaced by relational databases. The second is the object-oriented database. In this database, data is stored as objects that not only include characteristics, but also possible actions. The objects in an object-orientated database exist only when the database program is running on a computer (fortunately their characteristics can be stored for later use). Objects act on other objects, receiving and sending messages and processing data. For example, an object-oriented database of sewers may consist of objects with information about the size of the sewer pipe, but also how much water can flow through the pipe in a minute, and what happens to excess water. Additionally, a sewer pipe may have attributes that can be passed on to other sewer pipe records in the database, or modified to reflect the characteristics of another sewer pipe. Some GIS already use object-oriented databases.

Relating database tables

FID	Shape*	NAME	STATE_NAME	STATE_FIPS	CNTY_FIPS	FIPS
0	Polygon	Lake of the Woods	Minnesota	27	077	27077
1	Polygon	Kittson	Minnesota	27	069	27069
2	Polygon	Roseau	Minnesota	27	135	27135
3	Polygon	Koochiching	Minnesota	27	071	27071
4	Polygon	St. Louis	Minnesota	27	137	27137
5	Polygon	Marshall	Minnesota	27	089	27089
6	Polygon	Beltrami	Minnesota	27	007	27007
7	Polygon	Lake	Minnesota	27	075	27075
8	Polygon	Polk	Minnesota	27	119	27119
9	Polygon	Pennington	Minnesota	27	113	27113
10	Polygon	Clearwater	Minnesota	27	029	27029
11	Polygon	Red Lake	Minnesota	27	125	27125
12	Polygon	Itasca	Minnesota	27	061	27061

FIPS	POP1990	POP2000	POP90_SQMI
27077	4076	4651	2
27089	5767	5088	5
27135	15026	16076	9
27071	16299	14760	5
27137	198213	193111	29
27089	10993	9937	6
27007	34384	39655	11
27075	10415	10817	5
27119	32498	30353	16
27113	13306	13569	22
27029	8309	8090	8
27125	4525	4155	11
27061	40863	44445	14
27107	7975	7468	9

FIGURE 6.2. Database table relates (or joins) are made by identifying the same attributes in the key fields from two separate tables.

Representing and Communicating

The database's role for storing geographic information makes it central to the process of communication, also for maps produced from geographic information. Representation and communication usually involve databases. Databases are part of the technologies we encounter daily and are a field of study, management area, and science in their own right. Most geographic information and maps only scratch the surface of what databases can be used for, but the two most common uses of databases for geographic information and maps are as follows:

- Databases store measurements and observations of things and events.
- Databases store the symbols, values, and other graphic elements that help maps communicate.

Graphics drawn automatically by computer software generally are less refined aesthetically compared to graphics humans make directly. Hand-drawn maps and graphics can be used to improve communication. This also becomes necessary if the computer-produced graphic should be revised or geographic information is unavailable.

The organization of the database tables, records, and fields is called a "data model." The creation of a data model is an important task and needs to be considered in conjunction with the geographic representation, cartographic representation, and communication objectives.

When working with geographic information or maps, you should be aware of how the database can constrain representing and communicating. This may be the result of using software or hardware that is not adequate to the task, or due to the misuse of the database. The relations between different database tables can lead to a variety of errors. A common example for roads is that records in one database table use initial capitals and full names for street designations (Road, Street, Lane, Avenue) and another database

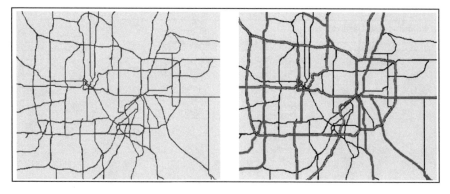

FIGURE 6.3. Highway and major road network in the Twin Cities, Minnesota, United States. The map on the left shows the roads without symbols.

FIGURE 6.4. Examples for symbols used in classifying roads in the United States.

table uses abbreviations (Rd., St., Ln., Ave.) or fails to use initial capitals. In most cases the database software will not relate the two tables because at the database's level of analysis (e.g., ASCII code) the street designations "Road" and "Rd." are different and unrelated.

Data Types

Database representation in most relational databases is restricted to a limited number of data formats. The terms used here have other analogous terms, some of which are indicated.

Integer: Whole numbers (1, 2, 3, 4,) are stored as integers.
Floating-point: Numbers with decimals (1.1, 2.4, 5.4,) are stored as floating point numbers.
Character string, fixed, or variable length: Also called "text," character strings can be usually stored only in fixed-length fields. In some cases the database software provides variable-length fields. If a word or text is longer than the fixed length of the field, the characters after that place will not be stored.

FIGURE 6.5. Examples of proper field length (top); too short field length (middle); too long field length (bottom).

Date and time, time interval: Because of the unique ways for storing date and time, most database systems offer separate data types for recording date, time, and, in some cases, time interval.

Simple large objects: Any kind of data (including images, word processor files, and spreadsheets) can be stored in a database as a simple large object as binary data. Because this data type is most often used to store binary data, especially images, it is often called Binary Large Objects (BLOBs).

An important issue to consider practically when creating a database is specifying the length of each field, also called "precision." If the field is too long, the database may require a great amount of computer storage space. If it is too short, attributes may be truncated (cut off) possibly making it impossible to know what the attribute actually records.

Data Storage and Applications

Considerations in designing a geographic representation or cartographic representation are the main factors determining how data is stored in a database. The available data types and the allocation of storage for attributes also play important roles. If the geographic information or map should show demographic characteristics of an area, most of the data will be stored in integer format. Data showing ratios will require floating-point fields. Text-type fields can be added for notation. For an application modeling erosion processes, the data will also be mainly numbers, but the types of observations and analysis will require mainly floating-point data types. Of course, if the geographic representation has led to personal addresses stored in a single field as a character string, it will be very difficult, possibly even impossible, to identify only those people in the database that live on Main Street.

The application type should guide practical considerations of which data types should be used in analysis and communication. For many purposes, observations and data recorded in numerical formats are the most flexible. They can be transformed and analyzed with other numerical data. Character strings are useful for recording the names and designations of things and events; BLOBS are usually used for images; data and time data types are used to record when things were recorded or events took place.

Entities and Relationships

A key part of working with databases is creating a data model that accurately and correctly shows things or events and their relationships. The clarity of this data model is important for others who need to understand the geographic representation and cartographic representation, or perhaps just the data model. A data model should describe each entity and the attributes that are associated with the entity's key identifier. Relationships are based on key identifiers and can be between two unique entities, one unique identity to several other entities, or between several entities.

A variety of techniques have been developed to sketch and make schemes showing the data model. Usually these techniques follow the entity-relationship conceptual understanding of a relational database. Entity-relationship diagrams, or E-R models, are often made using a graphical form based on the Universal Modeling Language (UML), which offers a systematic way of going from the diagram to conceptual and actual database description.

The capability of creating relations between data is extremely powerful and useful. However, relations work only when data is stored using the same format. For instance, returning to the address example from above, if the entire address is stored as one database field, it will require additional processing to relate this data with address data separated into multiple fields.

If the data can be related, the relation can be permanent or temporary. Any database processing of a relation that produces a single, permanent new table is called a "join"; otherwise it is just a relationship.

Normalization

Data normalization is a process of assuring that a database can take best advantage of relational database principles. If you normalize a database you can not only improve its performance, but avoid some organizational and logical errors that could diminish the quality of the database. Data normalization of relationship database technology was first described by Edgar Codd in the 1970s. The first level of data normalization requires that each field contain only one value (e.g., only the house number, not the house

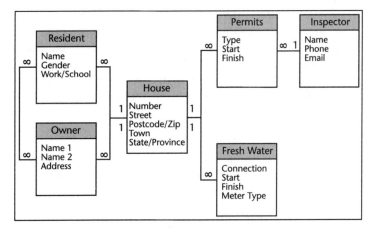

FIGURE 6.6. Entity-relationship (E-R) diagram.

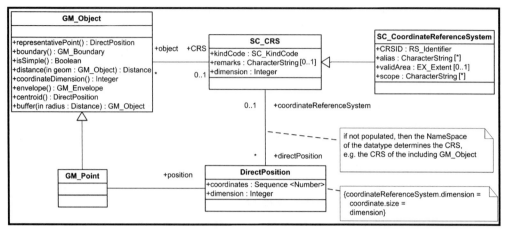

FIGURE 6.7. UML diagram for a point object following the ISO UML standard.
Courtesy of Michael Lutz.

number and street name). The second level requires that each value of a record is dependent on the key value of the record (e.g., the name of the person). In the third level, no fields depend on nonkey fields (e.g., a "years at residence" field must be related to the name of the addressed person, not the street number).

Data Modeling, Geographic Representation, and Cartographic Representation

The consideration, inclusion, and representation of the complex spatiotemporal relationships in a database pose a number of challenges that require a thoroughgoing engagement with geographic representation and cartographic representation. If David Sinton's matrix (see Chapter 2) provides a means to conceptualize the observation, measurement, and storage of data from a single geographic thing or event, the data modeling for a database must consider multiple things and events as well as the relationships. In addition to these issues of geographic representation, data modeling takes cartographic representation into account in varying degrees. The type of media, the projection, the coordinate system, and symbolization all influence data-modeling decisions. Relational database have several advantages for flexibly and reliability in addressing these issues. The relations between tables can reflect different relationships between things and events and multiple relationships, representations, and types of communication can be part of the data model. Of course, challenges exist when developing new geographic information when some geographic information already exists and when geographic information from different sources should be combined.

Summary

Databases should consider both geographic representation and cartographic representation. Based on the nature of the intended geographic information uses, the creation of a database could consider only geographic representation.

GIS databases are in almost all cases relational databases, which have great flexibility for geographic information and cartographic needs. You can think of a database as a set of tables that can be put into relationship with each other based on characteristics, or what are usually called "attributes." They are stored as one of several data types: integer, floating-point, character, data and time, or simple large binary objects. These tables are different from spreadsheets because each value of a characteristic is kept grouped into a record of all recorded characteristics for that database entity or object. Tables in relational databases are related using the entity-relationship model. These database tables are also used for recording attributes that are used for the symbolization of things and events. Data modeling plays a key part in preparing the geographic representation and cartographic representation of geographic information or maps.

In-Depth ASCII Characters

The partial list of ASCII (American Standard Code for Information Interchange) characters below contains the characters and their decimal representations, which is how the computer stores them. These numbers are further encoded by the computer in the hexadecimal, and then the binary format. First, you may note that ASCII characters also include nonprintable characters (some are not shown in this extract of the character set). You should also note that letters from other alphabets (e.g., ü or ą) are missing. ASCII is being replaced by the character set UNICODE, which includes those letters, as well as characters and symbols from many, many other languages. It will be some time, though, before this newer character set is fully phased in.

Dec. Val. Char.					
	48 0	65 A	82 R	99 c	116 t
32 space	49 1	66 B	83 S	100 d	117 u
33 !	50 2	67 C	84 T	101 e	118 v
34 "	51 3	68 D	85 U	102 f	119 w
35 #	52 4	69 E	86 V	103 g	120 x
36 $	53 5	70 F	87 w	104 h	121 y
37 %	54 6	71 G	88 X	105 i	122 z
38 &	55 7	72 H	89 Y	106 j	123 {
39 `	56 8	73 I	90 Z	107 k	124	
40 (	57 9	74 J	91 [	108 l	125 }
41)	58 :	75 K	92 \	109 m	126 ~
42 *	59 ;	76 L	93]	110 n	127 DEL
43 +	60	77 M	94 ^	111 o	
44 ,	61 =	78 N	95 _	112 p	
45 -	62	79 O	96 `	113 q	
46 .	63 ?	80 P	97 a	114 r	
47 /	64 @	81 Q	98 b	115 s	

Note: More complete ASCII tables are available on the Internet.

Review Questions

1. What is a database "join" operation?
2. What are the common field types used to store data?
3. What is a flat-file database?
4. What is the relationship between database representation and geographic representation?
5. What is the difference between *logical* and *symbolic* representation?
6. What is tabular information?
7. What is a relational database?
8. What does "database normalization" refer to?
9. What is an entity-relationship diagram?
10. How are cartographic symbols stored in databases?

Answers

1. What is a database "join" operation?

   A "join" operation permanently links two database tables based on a common record value.

2. What are the common field types used to store data?

   Common field types are character, integer, real, binary, exponential, and image.

3. What is a flat-file database?

   A flat-file database stores values in rows and attributes by columns. It is one way of representing entities in a database.

4. What is the relationship between database representation and geographic representation?

   Database representation is how items are symbolically stored and manipulated in a database. It is based on a geographic representation. For example, a road, geographically represented as two lines, with each line with attributes indicating the number of lanes in a constant direction, can be represented in another database as a single geometric line with two values indicating the number of lanes in each direction.

5. What is the difference between *logical* and *symbolic* representation?

   Logical representation is how *symbolic* representation is systematically recorded and stored in a database.

6. What is tabular information?

   Tabular information is database data represented in a meaningful tabular form.

7. What is a relational database?

   A relational database is a database system that allows for the association of data from different tables.

8. What does "database normalization" refer to?

   Normalization is the process of systematizing all relations among database elements and tables for consistent storage and efficient access of data. This reduces data redundancy and improves software and hardware operation.

9. What is an entity-relationship diagram?

   A figure to show the conceptual model of a database including all entities and relationships.

10. How are cartographic symbols stored in databases?

    Each software package uses its own specific solution, but generally a table of graphic symbols in the software is associated with graphic commands to draw the symbols and related to individual values in a database.

Chapter Readings

Martin, J. (1983). *Managing the Data-Base Environment*. Englewood Cliffs, NJ: Prentice Hall.

Rigaux, P., & M. Scholl, et al. (2002). *Introduction to Spatial Databases: Applications to GIS*. San Francisco: Morgan Kaufmann.

Shekhar, S., & S. Chawla. (2003). *Spatial Databases: A Tour*. Upper Saddle River, NJ: Pearson.

Web Resources

Note: You can find thousands of excellent descriptions of databases on the web. They range from those that are very technically advanced to those that are very simplistic. A search on the words "database introduction" should give you a list of many from which you can quickly choose the introduction best-suited to your needs.

Shashi Shekhar maintains a website with a few chapters online from his book introducing spatial databases (highly recommended, especially for people with some programming or IT experience). See *www.cs.umn.edu/research/shashi-group/Book/*

A short introduction well suited to people with only a little computer experience is available at *www.awtrey.com/tutorials/dbeweb/database.php*

For an introduction to SQL, try the interactive tutorial at *http://sqlzoo.net/*

Exercises

1. Normalizing a Database

Give students a nonnormalized database of cartographic symbols and features. Have them normalize the database by using a provided symbol table.

2. Create a Theoretical Model of a Geographic Relational Database

Objective

Learn principles of databases, geographic representation, and cartographic representation.

Activities

Students should work from a partial model related to their interests or field of studies to add entities related to a particular application to geographic data.

Chapter 7

Surveying, GPS, Digitization

Collecting and communicating reliable geographic information about things and events requires knowing in a systematic fashion where they occur. Projections, location systems, and coordinate systems provide key geographic reference frameworks for systematically locating observations and measurements of distinct locations. Geographic representation and cartographic representation rely on the collected location information to make accurate and reliable GI. With various techniques of recording location, surveying, GPS, and digitalization are three generic ways of recording the locations and characteristics of things and events by directly observing them or indirectly measuring their location. In all three forms of location measurement, the collection of positional information requires the systematic collection of measurements.

Geography distinguishes between position and place, though the terms in many other usages are often synonymous. "Position" refers to the systematic measurement of the place associated with a thing or event. "Place" only refers generically to the site, usually referring to something more familiar. For example, the *place* where the Eiffel Tower is located is Paris, France. The *position* of the Eiffel Tower is approximately East 2.37 longitude and North 48.7 latitude. We should also note that we use the word "approximate" when referring to position unless we have accurate measurements of location in a geographic reference framework

Considerations of geographic representation and cartographic representation issues have strong impacts on the methods and techniques used for collecting positional and attribute data. How we wish to show something by itself and in relationship to other things and events involves defining a number of characteristics which in turn specify how the position of objects is determined. For instance, the location of a forest may be known reliably, but the location of the trees in the forest is another matter. If we need to know the location of the trees we have to decide what a tree is (to avoid including bushes, no matter how large) and where a tree is. For example, these three options for surveying the location of trees could be considered:

- The tree is located at the center of its trunk at breast height.
- The tree is the located at the northern-most point of the stem, not including surface roots.
- The tree is located at the center of its canopy, that is, the maximum reach of its branches and leaves.

Depending on our purpose and how we want to geographically and cartographically represent the tree, one of these three approaches or perhaps a different approach would be used. If we are conducting an environmental analysis, knowing tree locations with an accuracy of 1–2 feet or even meters may be sufficient. If we need to be more accurate, as accurate as a fraction of an inch or a few centimeters, we need to rely on the help and services of an expert surveyor, or a geodesist as they are also called.

The discipline of surveying and geodesy specifies methods, techniques, and procedures when high accuracy is required for legal, building, or other purposes. Surveyors are often called upon to meet legal requirements, but they could also be to satisfy our desire to know as accurately as possible where things are.

GPS is commonly used for recording locations for a variety of applications. It usually works less well in forests and where there are other obstructions to the signals (for reasons discussed below), but can still be used. If costs are a major issue—for any number of reasons—digitization of existing materials may be a viable option, provided they are available and this form of use is permitted. Many copyrights on maps prohibit using them as the basis for digitization. Assuming that the maps are available and not copyrighted, digitizing GI can be a good compromise and a reasonable way to collect positional information.

Surveying

With the increasing use of specialized technologies, surveying has become a complex field, but very basic surveying techniques for hobby or curiosity can be practiced by most people. These techniques are elementary and the process requires a minimal amount of mathematics and geometry. The emphasis in this presentation is on the broad understanding of surveying, but this section will also lay out some of key issues for advanced legally and disciplinary regulated surveying. More and more people survey, which makes it ever more important to know what surveying is and why and when regulations and licenses of surveyors are necessary.

What Is Surveying?

Surveying, broadly understood, is the field collection of positional and attribute information using direct and indirect measurements. More narrowly understood, as a discipline, surveying is the regulated methods, techniques, and procedures of position determination for legally regulated activities, engineering, and other activities requiring certifiable accuracy. Surveying is also known as geodesy in many areas, especially when very recent technologies have become a mainstay of the surveying.

We can define surveying as the systematic collection of positional location and other location-related characteristics. It is an organized activity using known coordinate systems and procedures for attribute collection based on geographic representation and cartographic representation. The collection of positional and attribute information in the field must resolve the problems of reducing measurements from the infinitely complex earth to observations that correspond to the geographic representation and cartographic representation.

Advanced surveying to fulfill the needs of construction and legal requirements is a very specialized discipline. Technologies and methods define the practices of surveying; laws and regulations define the standards and practices.

Brief History of Surveying

Even without telescopes, tape measures, or lasers, ancient surveyors could do work of astonishing accuracy. The pyramids in ancient Egypt are evidence of that accuracy which exemplifies the advancement of Egyptian surveying. Even older map fragments found in Mesopotamia (modern Iraq) point to that society's advanced surveying techniques. Even if we can only puzzle over the construction of neolithic monuments in Stonehenge, Easter Island, and

FIGURE 7.1. Professional surveyors need high-accuracy equipment such as this prism pole, which reflects the laser light used in detailed surveys.

Photo courtesy of Crain Inc.

other places, these monumental works' locational accuracy in reference to movements of the solar system demonstrates their creators' surveying skills.

For most of known history, surveying has been a very stable discipline, only changing as new instrument-making technology advanced and survey accuracy increased. If a surveyor from ancient Egypt had been able to travel through time and go to any Western country up until the 1880s, he or she would have found the accuracy of survey measurements greatly advanced, but the techniques and basic instruments remarkably similar. Surveyors used a chain of fixed length as a common instrument to measure distances for surveys in many parts of the United States until the 20th century.

During these four millennia, surveying involved numerous techniques that can be simplified first into distance and angle measures and second into leveling. Distance and angle measure involved the use of a plane table to make situation drawings "in the field" and devices such as telescopes to make accurate measures. Leveling was done with plumb bobs and water and mercury levels to accurately measure changes in elevation between locations and their respective heights. Surveying was often connected to navigation and most surveyors were capable of navigation by sextant using the stars or sun.

Basic Field Survey Techniques

The most elementary techniques for collecting positional information only require instruments for measuring distances and angles. Usually collected by making drawings on a plane table (something like a breadboard kept level on a tripod), a basic field survey starts at a point with a known position. Measurements taken for a survey continue through a series of distance and angle

---

In-Depth **The Gunther Chain**

One of the unsung technical heroes of American history is certainly the Gunther chain. The Gunther chain greatly simplified surveying of public land, thereby easing the development of the western United States.

Gunther's chain was a measuring instrument 66 ft (20.1 m) long, divided into 100 links (1 rod or perch = 25 links). Each link was a short section of wire connected to the next link by a loop. It was long used for land surveying and became a unit of length (80 chains = 1 mile). These units are still used in many surveys.

Although its inventor Edmund Gunther invented this measuring device in the late 16th century, it wasn't until the late 17th century that it was widely adopted. Once adopted, it was the most common unit of measurement for U.S. surveyors until the early 20th century. Because it was hand-made, each chain was somewhat different in length and surveyors needed to use a table of adjustments for this distortion and for distortions arising from heat and cold. A surveyor also needed another person, a chain man, to help carry the chain and look after it.

FIGURE 7.2. Surveyors at work with a plane table.
From *http://erg.usgs.gov/isb/pubs/booklets/topo/topo.html*

measurements which are verified against each other using trigonometric equations. At the same time changes in elevation are recorded.

A survey of positions collected in this manner may be accurate by itself, but it could not easily be combined with other surveys and other GI to make maps. Lacking a clear relationship of at least one point (four are for statistical reasons the practical minimum to consider) to a vertical and horizontal datum, it would be very hard to connect the surveyed positions to any coordinate systems.

More advanced survey techniques rely on defined procedures and rules. These techniques are a basic part of a trained surveyor's skills because of much greater error control and accuracy measurements than possible with basic field survey techniques.

FIGURE 7.3. Geodetic markers are part of national triangulation networks and are connected to geodetic datums.

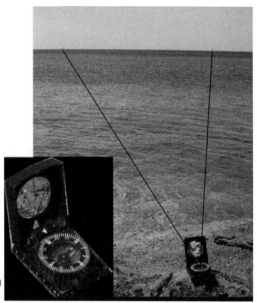

FIGURE 7.4. A compass can be used for navigation and surveying.

GPS and GNSS

Many people have heard of GPS, or global positioning systems, many more have used it without even knowing it. A common feature in new cellphones, cars, and boats, GPS has been used for years by a wide and varied group of users including trucking companies, buses, hikers, taxis, surveyors, and air lines. GPS has become commonplace because of its ease of use and accuracy in determining location. While its level of accuracy is insufficient for professional surveyors, for most people, most of the time, GPS is accurate enough. Car navigation systems, which are becoming very common, offer the ability to show where the car currently is and to get instructions about how to get to another place. The instructions can be shown on a display with a map or spo-

FIGURE 7.5. Basic concepts of leveling for a survey.

FIGURE 7.6. GPS navigation system built in as an integral part of a car's dashboard.

ken, making it easier for the driver to remain concentrated on his or her driving. The navigation systems generally work well, but can become a bit nagging when the computerized voice incessantly rattles off the changing names of streets and twists in a road. Other applications—for example, navigation systems for visually impaired people—point to the many potentials.

Because of its ubiquity and importance for so many different activities, the term GPS, which refers to only the U.S. funded satellite-based position-finding system, is slowly being replaced by the term "global navigation satellite system" (GNSS), which is the broader term. Because the only other GNSS are currently under development (Galileo) or of limited and specialized use (GLONASS), this book uses the term GPS, although in a few years, as these other systems become more operational, GNSS will certainly find wide usage.

WHAT IS GPS?

The Global Positioning System (GPS) is a system of satellites launched and maintained by the U.S. Department of Defense. (They refer to GPS as

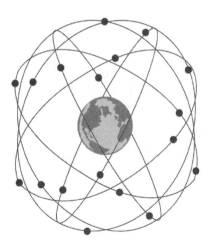

FIGURE 7.7. Idealized drawing of GPS satellites in orbit.

From Campbell (2007), p. 379. Reprinted by permission of Guilford Publications.

FIGURE 7.8. The only GPS satellite on public display.
From *http://commons.wikimedia.org/wiki/Image:Global_Positioning_System_satellite.jpg*

NAVSTAR GPS.) Over 50 satellites have been launched, each about the size of a school bus. The system costs around $400 million yearly to maintain, but is freely available all around the world. The accuracy and availability can be limited by the Department of Defense if they see a need through selective availability, which degrades the signals received by commercial GPS units.

With a GPS receiver, which can be a computer chip close to the size of a postage stamp attached to an antenna, a device can receive and process the GPS satellite signals and determine location and elevation. How this works is quite complex, but the general idea is rather straightforward. Instead of using a measurement of distance, as in surveying, GPS uses the time it takes radio signals to travel. With much simplification we can say that the GPS receiver calculates the difference between its own clock's time and the time communicated in signals from GPS satellites, then uses this difference to calculate the distance between the receiver and the satellites. The time difference is detected in the difference between the signal sequence (a binary signal called "pseudorandom") received by the GPS receiver and the signal sequence it has. Each satellite broadcasts a signal that contains data about the satellite and the time on its clock. The GPS receiver's time calculations also should take a variety of interferences into account, especially interference in the atmosphere of the earth that can slow down the transmission of a signal from a satellite. If the GPS receiver is traveling, the corresponding movement must also be taken into account.

Obviously some limits to the accuracy of the GPS measurements arise, related to the number of satellites available. The most complex part of GPS positioning is the determination of location. With the signal from one GPS satellite, a GPS receiver can only determine how far it is from that satellite, but not where. It could be anywhere on an imaginary sphere drawn at that distance from the satellite. To determine the position of the GPS receiver, signals from at least four GPS satellites are needed if no other information is available. If the elevation of the GPS receiver is known, only two satellites are needed to determine position.

With the variability of the atmosphere, the movement of the GPS receiver, and possible obstructions in the local environment of the GPS receiver, the accuracy of GPS positioning may be limited. The factors that reduce the accuracy are summarized in the measurement called the positional dilution of precision (PDOP). Larger values indicate less accurate GPS positioning. If the values are greater than 8, than the positional location provided by the receiver is very inaccurate. Values less than 4 are a good indication of high accuracy.

Various other factors impact the positional accuracy of GPS measures. The accuracy of most GPS receivers is less than 3 m under ideal conditions. The accuracy is often even around 1 m. If the PDOP value is less than 4, the positional values may even be accurate down to 1 or 2 m. Most GPS receivers take atmospheric interferences into account by using information about the atmosphere at a given time. More accurate receivers compare the speeds of the GPS satellite signals to calculate the reduction in positional accuracy. Most receivers also take the reflection of GPS satellite signals off of the ground and buildings (multipath error) and changes in satellite orbits into account. A receiver can reduce error by choosing satellite signals based on the characteristics that introduce error into the position locations and using satellites that help produce the mathematically most accurate results. The

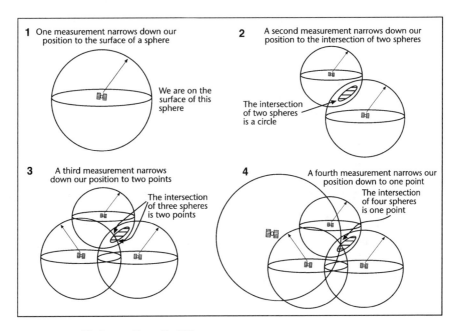

FIGURE 7.9. Finding position with GPS.

quality of the receiver also contributes to accuracy. Higher quality receivers have more or better procedures for reducing errors.

Greater accuracy can also be achieved through a variety of techniques, regardless of the cost of the receiver. The most common of these (and the most widespread) is called the differential global positioning system (DGPS; see Plate 4). DGPS relies on a fixed GPS receiver that can calculate GPS changes in its position against the accurate surveyed position. The difference indicates the current amount of error in the GPS system for a particular area. The fixed location DGPS receiver broadcasts a correction signal used by GPS receivers to adjust the GPS signals and get highly accurate (down to centimeter or inch accuracy) positions. The most common DGPS is called the wide area augmentation system (WAAS) and is available in North America.

APPLICATIONS

Most people who have used a GPS probably can't imagine any limit to its applications. Even if its shortcomings are grievous (it can't be used indoors, nor very well in a forest, or where there are many tall buildings or cliffs), solutions have been developed to these problems. Usually these solutions involve broadcasting radio signals or pseudo-GPS signals that are highly accurate. The configuration of these systems is very complicated and requires large institutional investments. Most are made by governments. For instance, the European Union is developing a high-accuracy network (along the lines of the U.S. WAAS) for navigation purposes. Even if a GPS receiver lacks the ability to use these extra networks, GPS can still be used in a number of applications, some of which are described here.

Vehicle Navigation Systems. More and more people have used GPS-based systems in cars; many more have benefited from the use of GPS in cars, buses, trains, and trucks. The GPS receiver may be hidden in the dashboard, but may be critical for the taxi company to find out which taxi is closest to you when you call for a pickup. A GPS receiver can help a trucking company better organize deliveries to minimize the fuel used. A bus may have a GPS installed to help the bus company indicate to passengers how long they need to wait for the next one.

Navigation systems are used for more than vehicles on land. They are also widely used for nautical and aeronautical navigation. They have become for many sailors irreplaceable because they work regardless of the weather and can easily be combined with computerized chart information. Almost all planes use, or will use, GPS. Together with high-precision positional transmission, planes can use GPS-based systems to land in any weather with centimeter precision.

Hiking. More and more hikers turn to GPS to help them find out more exactly where they are and to help them to plan a route before they go. GPS may not be reliable in canyons or along steep cliffs, but in most situations and weather it provides accurate positional information. Some map makers have started to change their map designs to make it easier for hikers to use. Some tourist areas offer GPS for people to help them follow a certain tour.

Aids for the Visually Impaired. Combined with acoustic or tactile signaling devices, GPS can be used to help visually impaired people find their way in new settings and navigate places that rapidly change—for example, a state fair or a college campus, as was done by Professor Reg Golledge and others at the University of California at Santa Barbara.

OUTLOOK: NEW SYSTEMS

The U.S. GPS system is widely and freely available, but because the U.S. Department of Defense controls it, people in other countries have little or no influence over its operation or when it might be shut down. The Commonwealth of Independent States (CIS) has a satellite system, which operates similarly to the U.S. GPS. But it has fewer satellites, offers less coverage, and is only available to people with very expensive receivers.

The European Union is currently developing a system, fully comparable to GPS, called "Galileo." Two prototype satellites were launched in late 2005 for testing. The full system, with 30 satellites and offering better coverage of polar regions, will be operational some years later.

Digitization

When maps exist, it is possible to convert them to GI using either tablet digitization, heads-up digitization, or a scanner. The reasons for digitizing from maps cover a gamut: the maps may be old and show something that people want to compare to recently collected GI, the maps may be unusual or hand-made, or the maps may be the only way of getting the desired GI.

Tablet digitizing involves the affixing of the source material (maps, drawings, etc.) which are georeferenced to coordinates on a table digitizer, a board of variable size. The location of the digitizer puck, the mechanical pointer calibrated to the digitizer and freely moveable, is recorded as different buttons are pressed. Software translates the location and button values. Heads-up digitizing is similar, but requires that digitized source material be georeferenced to a coordinate system. The material is displayed on a screen and the person doing the digitizing uses a mouse or similar pointer device and presses buttons to record locations. The scanning of existing map material is also common, and because it is mostly automated is very fast compared to tablet digitizing, especially when a large number of maps are to be scanned. But scanning usually requires complex postscanning cleanup. Hand digitizing is generally cheaper, but generally less accurate. Scanning is expensive for just a few maps, and may be complex to configure, but it gets cheaper if you have a number of maps prepared in the same way.

Keep in mind the following advice when preparing or working with surveyed, GPS'd, or digitized data:

ACCURACY AND PRECISION

The data collection procedures, tools, and techniques should assure the highest level of fidelity to the geographic representation and the cartographic representation. If the accuracy of the map materials is known, you have a great assistance in knowing how accurate the GI is. If not, it becomes complex. A rule of thumb is to always be more cautious than necessary when determining accuracy.

Of course, you have to be sure not to mix up accuracy and precision. "Accuracy" refers to the agreement between the GI or map position and the ground position, whereas "precision" refers to the number of digits used to indicate the position. High precision is meaningless without corresponding accuracy.

CHOICE OF POSITIONAL COLLECTION TECHNOLOGY AND APPROACH

The purpose of and means available for data collection largely will determine the collection technology. The most important additional factor here is often cost. If the data collection has to choose between two methods, generally the lower-cost option will win. The exception would be if the higher-cost option offers additional information, accuracy, or reliability. Of course, the lowest cost option can easily end up being the most costly in the end. Far too often, people collect GI without thinking through the geographic representation and cartographic representation.

CLOSURE OF AREAS AND CONNECTIONS BETWEEN LINES

A complicated issue for digitization is making sure that an area is closed (e.g., a county, state, or country) or a line is connected (e.g., a highway, bus route, or subway). This can be remedied by using a tolerance that moves digitized points together if they fall within a specified distance of each other. The tolerance is usually based on the accuracy of the GI. This can be difficult because the proximity of features changes across a map. Buildings are closer together in urban areas, some areas have long and narrow fields, others have very large rectangular fields. The tolerance for connecting points when digitizing needs to be adjusted to the circumstances.

GENERALIZATION EFFECTS

When digitizing maps, you need to bear in mind that generalization operations may have moved features on the map to make the map easier to read. This is common in small-scale maps, but also occurs in large-scale maps. If it is impossible to find out how features have been generalized, you can at least use the indicated accuracy of the map as an indicator of how much an individual feature could have been moved.

In examining positional collection technology options, you also need to consider remote sensing, discussed in Chapter 8.

PLATE 1. Undulations between the EGMS96 geoid and the WGS-84 geoid show differences in elevation around the world.

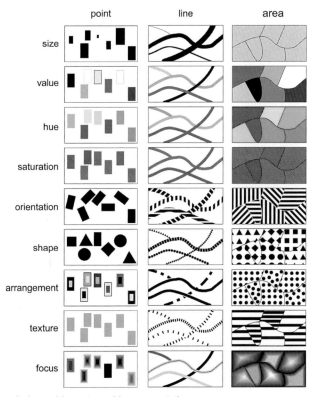

PLATE 2. Graphic symbols used in cartographic representation.

Source: MacEachren (1994). Adapted by permission of Alan MacEachren.

PLATE 3. Color value, color hue, and color saturation.

PLATE 4. Accuracy of GPS positioning with and without DGPS.

From *www.training.fema.gov/emiweb/downloads/DF109%20GPS%20IG.pdf*

Two Garmin Etrex GPS units were placed at the intersection of two sidewalks at the center of the cross hairs, with a clear view of the sky.

The yellow circle represents a 15-meter radius of the expected accuracy of a unit without DGPS. The yellow line represents the position given by the unit during one hour of recording. Notice that most of the error occurred when the unit was first turned on. After a minute, accuracy was within 5 meters.

The green circle represents a 3-meter radius of the expected accuracy of a unit with a DGPS signal. The green line represents the track. Once again most of the error (but still within 3 meters) occurred when the unit was first turned on.

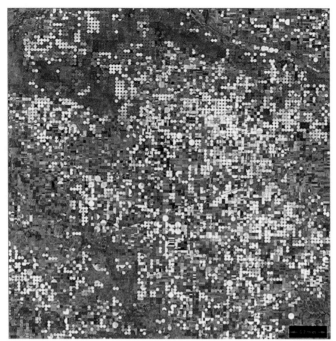

PLATE 5. Center-pivot irrigation systems create red circles of healthy vegetation in this image of croplands near Garden City, Kansas.

From *http://earthasart.gsfc.nasa.gov/images/garden_hires.jpg*

PLATE 6. The London Tube Map became the reference for easy-to-read maps of transportation networks in part because it only shows topological connections.

Reprinted by permission of Transport for London

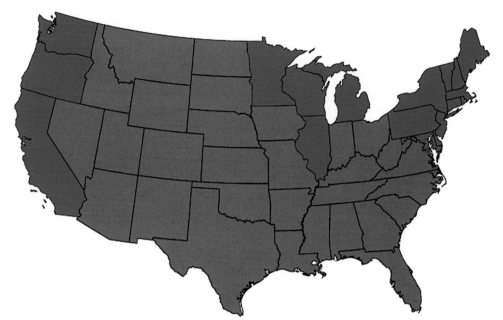

PLATE 7. Red states/blue states map showing U.S. presidential election results from 2004.
From *www-personal.umich.edu/~ mejn/election/*

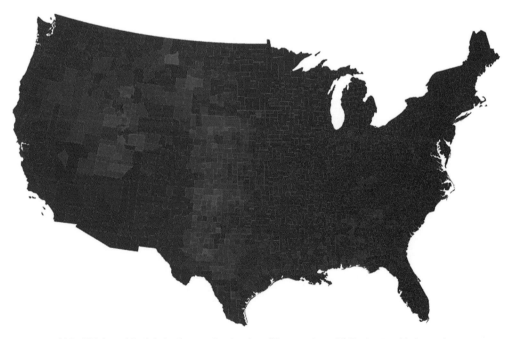

PLATE 8. U.S. 2004 presidential election results showing difference from 50% of vote with increasing saturation.
From *www-personal.umich.edu/~ mejn/election/*

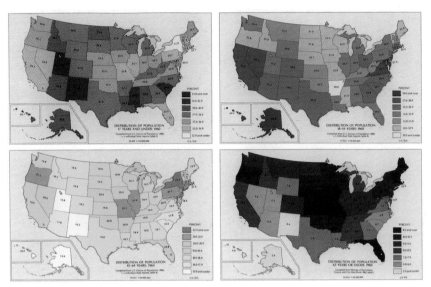

PLATE 9. Thematic maps are often used to create easily read comparisons.

PLATE 10. Cartogram examples of 2004 US presidential election results.

From www-personal.umich.edu/~mejn/election/

PLATE 11. Illustration of differences between choropleth (top) and dasymetric (bottom) mapping. From *http://geography.wr.usgs.gov/science/dasymetric.html*

PLATE 12. Precipitation amounts based on a field model generated using inverse distance weighting.
Courtesy of Chris Lloyd.

PLATE 13. Kriging (using the same precipitation data used in Plate 12).
Courtesy of Chris Lloyd.

PLATE 14. A bicycle ride across the United States provides the subject matter for adding to Google Earth's rich selection of geographic information. A visualization tool for most people, it has become emblematic for the new directions that GIS is going in the eyes of many users.

Image courtesy of Jason Dykes. Reprinted by permission of Google Earth™ mapping service.

PLATE 15. Jason Dykes maintains a spatial diary showing his travels in the United Kingdom. While Jason collects this data himself and controls it, a truck driver or rental car driver might be tracked unawares in a similar fashion.

Image courtesy of Jason Dykes. Reprinted by permission of Google Earth™ mapping service.

Summary

The GI for maps and other communication should be collected to fulfill requirements arising from issues related to the geographic and cartographic representation. Accurately and reliably locating observations and measurements about things and events requires careful consideration of the options for collection and the issues each option faces. The three generic options are surveying, GPS, and digitization. Traditionally, surveying was the discipline called on for accurate and reliable measurements of position. GPS, which is widely available today and becoming more commonplace, is altering that somewhat, but surveying remains the discipline called on for accurate and reliable location measurements, especially when legal dimensions of the things or events are important. Existing materials can be digitized or scanned, but copyright regulations and limitations should be carefully considered.

In-Depth Copyright Issues for Geographic Information

Almost all GI is protected implicitly or explicitly by copyright. The only blanket exception is for GI collected by the U.S. federal government. This, however, applies only to civilian agencies. Military agencies (including the Corps of Engineers) are exempt. Individual states have their own laws regulating the use of copyright for their agencies and other government agencies (towns, cities, etc.) in that state. Private companies have copyright on the GI—for example, a map made by a surveyor, unless otherwise defined or regulated.

Copyright sets out to motivate the expression of ideas by offering restrictions on how original works in a tangible medium may be used by someone else. In GIS, as the geographic representation and the particular cartographic representation are these works.

Charging for GI is commonplace all over the world. Copyright is a way to ensure that people who use GI created by others compensate them for their work. When a person uses copyrighted material, he or she has to request permission and/or reimburse the owner. This can get very expensive and very lucrative, so there are many people struggling over copyright and seeking exemptions that allow them to do what they would like to do without compensating the owner. The U.S. government took the stance over 200 years ago that copyrighting material created for and by the government would hinder commerce and be an imposition for the development of the economy. In considering the use of GI around the world, the results are clear: the U.S. GI economic sector is vibrant and there is widespread (even global) access to U.S. federal government GI. U.S. state governments sometimes take a different view. Some allow free access, some charge. Usually the laws and regulations offer a variety of exceptions, but some states have decided that because GI costs money to produce, users should be charged.

The laws of states regulating access to GI include open records laws, which are related to the Freedom of Information Act. In the United States these laws regulate access to GI (and other types of information created by the government). In Europe recently the Freedom of Information Act has led to a number of attempts to acquire GI from government agencies. The INSPIRE project should also increase the availability of geographic information. This is still a cumbersome process and may not lead to the desired results if parallels to the open records laws can be made. For example, digital access in an open record law may be sufficient (legally) if people can come to a government office and sit down at a computer and access the data. This is access, but not necessarily the access to the original GI on a different computer that most people might expect. In this, and similar, ways copyright often becomes a way to protect resources, rather than motivate the creation and dissemination of ideas.

Review Questions

1. What specific steps does the systematic collection of positional location entail?
2. Under what circumstances is surveying legally regulated?
3. What instruments are commonly used today for surveying?
4. What instruments are used traditionally for surveying?
5. What legal issue must be considered before using existing GI?
6. Where is the use of GPS less accurate?
7. Which accuracies does GPS support?
8. How should geographic representation and cartographic representation be taken into account for data collection?
9. How can generalized GI and maps affect positional accuracy?
10. What are common sources for existing GI?

Answers

1. What does the systematic collection of positional location involve?

   The systematic collection of positional location and other attributes is an organized activity using known coordinate systems and procedures for attribute collection based on geographic representation and cartographic representation.

2. Under what circumstances is surveying legally regulated?

   Generally surveying is legally regulated when it involves the collection of location information used for purposes or activities with possible immediate public safety consequences.

3. What instruments are commonly used today for surveying?

   Total stations, GPS (GNSS), and laser range finders are among the most common.

4. What instruments are used traditionally for surveying?

   The theodelite, measuring tapes, rods, and plane tables were traditionally used.

5. Can existing GI be used in any way?

The copyright status and distribution rights must be assessed before using GI from other sources.

6. Where is the use of GPS less accurate?

Generally, GPS is less accurate under tree foliage, near trees, near cliffs, or near high buildings, all of which can obstruct the GPS satellite signals.

7. Which accuracies does GPS support?

GPS (or other GNSS) is suited for any activities where information about location at a modest accuracy ($\geq$ 3–5 m) is needed. For higher accuracies, additional procedures, tools, and techniques can be used.

8. How should geographic representation and cartographic representation be taken into account for data collection?

The data collection procedures, tools, and techniques should assure the highest level of fidelity to the geographic representation and the cartographic representation.

9. How can generalized GI and maps affect positional accuracy?

Generalization distorts GI and maps in a variety of ways. This distortion reduces positional accuracy.

10. What are common sources for existing GI?

Many government agencies, national mapping agencies, and private companies are potential sources of existing GI.

Chapter Readings

Campbell, James B. (2007). *Introduction to Remote Sensing* (4th ed.). New York: Guilford Press.

The mathematical basis for geographic surveying is covered in:

Cotter, C. H. (1966). *The Astronomical and Mathematical Foundations of Geography*. New York: Elsevier.

For an older, but lucidly presented, text, see:

Hinks, A. R. (1947). *Maps and Survey*. Cambridge, UK: Cambridge University Press.

For a more recent text for students starting with geographic field work, see:

Lounsbury, J. F., & F. T. Aldrich. (1986). *Introduction to Geographic Field Methods and Techniques*. Columbus, OH: Merrill.

For a history of surveying and map making, see:

Wilford, J. N. (2001). *The Mapmakers*. New York: Knopf.

Web Resources

A good glossary of GPS and GNSS terms is available online at *www.magellangps.com/en/about/aboutgps/glossary.asp*

More on the surveying of the United States is available online at *www.measuringamerica.com/home.php*

An excellent starting point for resources on the history of surveying is available online at *www.fig.net/hsm/*

An overview of geodesy and its different uses is available online at *www.ngs.noaa.gov/PUBS_LIB/Geodesy4Layman/toc.htm*

A very thorough overview of the GPS system focusing on technical aspects is available online at *http://en.wikipedia.org/wiki/GPS*

Trimble, a large GPS hardware and software provider, offers an animated GPS tutorial is available online at *www.trimble.com/gps/index.html*

An overview and details about the roles of GPS for aeronautics (and beyond aviation) is available online at *http://gps.faa.gov/index.htm*

For information related to European GNSS activities, see *www.esa.int/esaNA/index.html*

The American Congress of Surveying and Mapping (ACSM) offers much information about current surveying training and activities online at *www.acsm.net/*

Exercises

1. GPS and Navigation: Good for What?

What are the advantages of using GPS for navigation? What kinds of navigation benefit the most? Why? What are some of the disadvantages? Are there potentials to abuse GPS and collect personal information about people?

Finally, what about the use of GPS for surveying? Do you think any person should be allowed to use GPS for surveying?

2. EXTENDED EXERCISE: Basic Surveying

Objective

Learn basic concepts of surveying, especially triangulation.

Overview

Most people don't think about it, but without surveyors, not much would happen in a modern world. Houses, banks, roads, bridges, airports—all structures–rely on accurate surveying. Property ownership is also surveyed, and when it's not done properly expensive legal conflicts are usually unavoidable. The mathematical foundation of surveying is Euclidean geometry. This is the oldest geometry in the Western world and the one that approximates very well our actual experience of distance relationships.

In this exercise, you will learn a little about one of the oldest and most reliable surveying techniques, the *plane table survey*, by preparing a simple survey and doing some geometric evaluation. While the results of this exercise will not be accurate, the technique you will learn, when conducted with the appropriate instruments and robust procedures, will allow you to survey many things. And you will have a new insight into why mathematics is important for maps.

Concepts

Many technologies are used in surveying. The plane table survey is an incredibly simple technology for fairly accurate surveys. The technology usually uses a large $2' \times 2'$ board positioned over reference points. By using an accurate instrument for sighting points (one type is called an "alidade") and by keeping a consistent scale for measuring distances, a surveyor measures angles that accurately bisect position points. In other words, a traditional surveyor uses angles to determine distances and rarely measures distances directly. (Using computer-based surveying devices employing laser range finders, surveyors now are likely to measure distances.)

For this exercise, you will be using a simple piece of paper as a pseudo-plane table working indoors (making it possible to do this exercise any time of year). In comparison to working with a plane table, this diminishes the accuracy of the survey; however, the concepts and techniques remain the same. Using measurements of angles, you will be applying Euclidean geometry's *law of sines* to construct a locational survey of items in the classroom. Make sure to look at the law of sines example before starting with the exercise.

Exercise Steps and Questions

Preparation

You need to have a piece of paper (called a worksheet) for drawing your measurements and making the basic calculations, which you should place on a pad or spiral notebook. You will also need a straightedge ruler and a few colored pencils. Your instructor will provide you with a protractor that you will use to measure angles.

Your instructor will have created several baselines in the lab room. Each line is measured in centimeters. Each baseline forms a side of the triangles you will construct to locate objects in the room.

What to Do

First, form a group of four or five people. Prepare your lab instruments and get one protractor for each group. *Each person should complete a sheet indicating all measures, constructions, and calculated distances.*

You will be surveying and determining the angles from the baselines to *five objects* in the room. Example objects are:

- Thermostat
- Window levers
- Wall sprinklers
- Emergency lights
- Clock
- Light switches

Step 1

To start with your survey, go to one of the baselines. Two points are indicated, one labeled *A* and the other *C*. Draw a line to scale on your worksheet positioned to fit the surveyed elements and label the points. If your baseline is in the front of the room, put it at the bottom of the page. No matter where you are in the room, remember to always keep the orientation of your page. To figure the scale, set up a conversion ratio—for example, 1 inch on paper = 100 cm in the room.

Sighting and drawing line with Plane Table

Step 2

Put point *A* on your worksheet directly over the corresponding point *A* on a baseline in the room. *Accuracy here is very important.* Make sure to keep the paper stable after you have found the right position. Now, take your straightedge and point it from the point on your worksheet to the object you will survey—clock, thermostat, or the like. Make sure you are very accurate in drawing the line with one of your lighter-colored color pencils. Draw a straight line at least long enough to cover the distance in scale (you will get better at this with experience). Move to the second point (point *C*), reposition your paper so that point *C* on the paper coincides with point *C* on the ground, and repeat the sighting with your straightedge and drawing of a line. The two lines should meet, forming a triangle. Help the next person set up, checking to see if he or she is also following the correct procedure.

Step 3

Now, use the protractor to determine the angles of the triangle you have just drawn. The protractor has two degree indications. Negative angles run from left to right, positive angles run from right to left. Use the indications that correspond to the direction of the angle, or direction of the "base" of the angle—for example, if the base of the angle (also called the "initial side") points to the right, use the angle indicators that run from right to left. Write the angle measurements on your worksheet together with the figures.

You can now use the law of sines to determine the distances from the baseline to the surveyed object. You only need to calculate one distance, but calculating both distances will be helpful.

Repeat Steps 1, 2, and 3 for the other four objects, making sure to position your worksheet accurately and measure angles very carefully. Put all your measures and the results of your calculations down on your worksheet. Work together with other people in your group to make sure everybody has the same (or almost the same) measures.

Evaluation

When you have finished surveying and calculating distances, answer the following questions.

Question 1: Draw a line in another color connecting your surveyed objects on your worksheet. Does it look like a straight line approximating the wall? Compare your measurements and calculations for each of the five objects you surveyed. How accurate were your measurements and calculations? What is the difference between your calculated positions and measurements? What explains the difference?

Question 2: You surveyed in only two dimensions. Would adding a third dimension for height make your survey less accurate or more accurate? Why? What about for more precise surveying work in general? What is the name of the process a surveyor conducts to assure accurate height measurements?

3. EXTENDED EXERCISE: The Law of Sines and Euclidean Geometry

Introduction

In this exercise, you will be using Euclidean geometry, named after the ancient Greek mathematician Euclid who lived around 300 B.C. and who wrote 13 books about mathematics collectively called *Euclid's Elements*. It is the most established approach to codify perceptions of space and motion. Euclidean geometry is also called "classical geometry" because many other people contributed to it and added to it over the centuries. Euclid's geometry consists of 10 axioms for fundamental geometrical relationships, such as the sum of the angles in a triangle always equals 180°.

Even though Euclidean geometry is very old and physics has modified its applicability to certain phenomena that are better explained by Einstein's theory of relativity, quantum dynamics, and so on, Euclidean geometry is very important for many modern activities ranging from surveying to computer-aided design, computer vision, and robotics. If you have ever played, or seen, a new videogame and been amazed by the graphics, a large proportion of the math behind those graphics is based on Euclidean geometry.

The law of sines is one of the most fundamental parts of Euclidean geometry used by surveyors. It expresses the relationship between an angle and its opposite side. In right angle triangles, the sine is the relationship between the opposite side and the hypotenuse. In any triangle, the ratio of one side to its opposite angle is the same as the ratio of any other side to its opposite angle. Expressed mathematically:

$$\frac{a}{Sin\ A} = \frac{b}{Sin\ B} = \frac{c}{Sin\ C}$$

The law of sines is related to the law of cosines and also to the law of tangents. These are more complicated formulas for solving for the lengths of sides and size of unknown angles.

Using the Law of Sines in Surveying

The law of sines is used to solve the length of an unknown side when you know the length of one side and two angles. In this example, I go through the steps to find out the length of c in this figure.

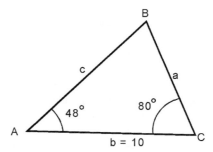

In the law of sines, all ratios are equivalent. If we know any three terms from two ratios, we can use basic algebra to solve for the unknown term. In this case:

$$\frac{a}{Sin\ A} = \frac{b}{Sin\ B} = \frac{c}{Sin\ C}$$

Now substituting the known terms

$$\frac{b}{Sin\ B} = \frac{c}{Sin\ C}$$

then using a sine table or the first three digits of a calculator's sine:

$$\frac{10}{.788} = \frac{c}{.985}$$

finally solving for c by multiplying both sides by .985

$$c = \frac{10(.985)}{.788} = 12.50$$

Additional Resources

If you want to find more information about the law of sines or Euclidean geometry, see:

Euclid's Elements (with interactive demonstrations) online at *http://aleph0.clarku. edu/~djoyce/java/elements/elements.html*.

Also see Geometry Reference Materials, available online at *http://mathforum.org/ geometry/geom.ref.html*.

Chapter 8

Remote Sensing

Remote sensing is the collection of data without directly measuring the object. It relies on the reflectance of natural or emitted electromagnetic radiation (EMR). EMR can be emitted by the sun (natural EMR) and, for example, sensed by photographic film, or it can be sent by a transmitter and the returned energy sensed, for example, by radar (a type of emitted EMR). Remote sensing has become a key means of data collection for a number of reasons, but mainly because it allows for systematic and accurate collection of GI.

Remote sensing is defined very broadly in this chapter as a measurement of an object's characteristics from a distance using reflected or emitted electromagnetic energy. This definition means that remote sensing includes all kinds of photography, aerial imagery, satellite sensors, and any kind of laser measurement. Remote sensing involves different types of sensor technologies ranging from photographic emulsions to digital chips. It also involves a vast array of storage media including everything from photographic film to computer files. As you can imagine, data from remote sensing and the sensor technologies themselves are a resource that can enhance the work being done in a number of other fields of study. For example, the discipline of surveying has changed enormously with the introduction of laser-based distance-finding technology.

The reason for defining remote sensing so broadly is that it is a very important GI technology. Remote sensing offers three advantages over other forms of data collection and GI. First, it makes it much easier to systematically recognize things and events over a large area. Second, it makes it easier and less costly to revise most maps. Third, digital remote sensing images can be used directly by other applications. There are some caveats to these advantages that you will find out about in this chapter. This chapter is purely introductory in nature and will skim over many of the crucial details and physics, but you should end up with a solid understanding of what remote

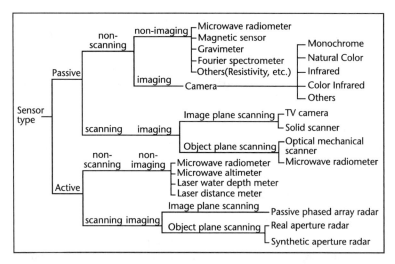

FIGURE 8.1. Different sensor types. Passive sensors use only reflected electromagnetic radiation (EMR). Active sensors use emitted EMR.

sensing involves and what some of the key issues and applications for remote sensing are.

Electromagnetic Radiation

Any understanding of remote sensing, regardless of the sensor technology, storage media, or application, starts with understanding EMR. First off, remote sensing's detection of EMR has three characteristics:

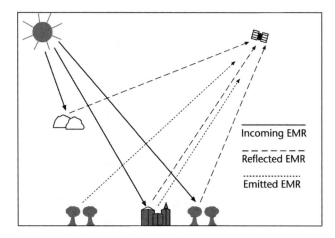

FIGURE 8.2. Emitted and reflected electromagnetic energy.

1. It generally only detects EMR from the surface of an object, although some sensors allow for penetration.
2. There is no contact between the sensor and the object.
3. All remote sensing measurements use reflected energy (usually from the sun) or emitted energy (e.g., from a radar station or plants).

The EMR detected by remote sensing technologies varies. It depends on the desired application as well as on the cost of different remote sensing data collections.

Spectral Signature

The EMR emitted or reflected by a thing or event varies. These differences are the basis for distinguishing things and events. The reflections and EMR emissions of a particular thing or event can be associated with a particular spectral signature that is used to identify where these things and events are located in a remote sensing image.

EMR also varies by time of day, season, weather conditions, moisture levels in the soil, wind, and a number of other factors. The physics involved in addressing these variations in emitted or reflected EMR is critical to the success of remote sensing and provides a commonplace solution. This solution, called "ground truthing," involves having some people in the field before, during, or after data collection who may take similar sensor measurements or observations. These measurements and observations can be used later to verify the remote sensing image or data and possibly to define correction parameters for adjusting the remotely sensed data to correspond to ground observations. Needless to say, this is highly complex and requires

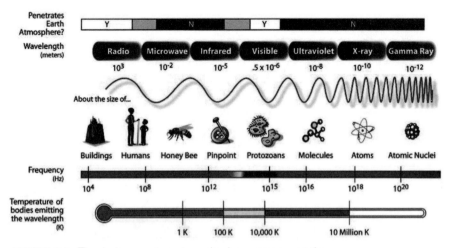

FIGURE 8.3. The electromagnet spectrum showing common examples.
From *http://lmynasadata.larc.nasa.gov/images/EM_Spectrum3-new.jpg*

FIGURE 8.4. Examples of spectral signatures. Note that a micrometer is one millionth of a meter. Source: *http://rst.gsfc.nasa.gov/Intro/Part2_5.html*

very-well-trained specialists to assess these factors and detect patterns in the remote sensing data.

Bands

The detection of patterns is helped by the use of different ranges, or bands, of EMR in sensing technology. Each band, as they are commonly called, refers to a particular range of wavelength for that sensor. The bands available for a particular sensor depend greatly on the purpose of the sensor and the technical characteristics of the sensor. Some sensors have only a few bands in a narrow range of the total EMR, others are much broader. For example, Landsat 7, the latest of the Landsat remote sensing satellites (discussed later in more detail), has seven bands:

Band 1: 0.45–0.52 µm Blue–Green
Band 2: 0.52–0.60 µm Green
Band 3: 0.63–0.69 µm Red
Band 4: 0.76–0.90 µm Near IR
Band 5: 1.55–1.75 µm Mid-IR
Band 6: 10.40–12.50 µm Thermal IR
Band 7: 2.08–2.35 µm Mid-IR

Figure 8.5 shows the different bands and how they can be combined for an application.

Another widely used satellite, SPOT 5, offers a different set of bandwidths.

FIGURE 8.5. Illustration of different bandwidths used by Landsat 7.

From *http://landsat.gsfc.nasa.gov/education/compositor/*

Resolution

Remote sensing distinguishes between spatial, temporal, and spectral resolution. *Spatial resolution* is the size of the unit recognized by the sensor, *temporal resolution* has to do with how often a satellite passes over and/or takes readings of the same spot, and *spectral resolution* measures the range of wavelengths the sensor can record. A raster cell is often also referred to as a "pixel."

TABLE 8.1. SPOT 5 EMR Spectrums and Bands

Electromagnetic spectrum	Spectral bands
Panchromatic	0.48–0.71 μm
Green	0.50–0.59 μm
Red	0.61–0.68 μm
Near infrared	0.78–0.89 μm
Midinfrared (MIR)	1.58–1.75 μm

TABLE 8.2. Comparison of TM and ETM+ Spectral Bandwidths for Landsat 5–TM and Landsat 7 (Source: *http://landsat.gsfc.nasa.gov/guides/LANDSAT-7_dataset.html*)

Sensor	Bandwidth (μ) Full Width–Half Maximum							
	Band 1	Band 2	Band 3	Band 4	Band 5	Band 6	Band 7	Band 8
TM	0.45–0.52	0.52–0.60	0.63–0.69	0.76–0.90	1.55–1.75	10.4–12.5	2.08–2.35	N/A
ETM+	0.45–0.52	0.53–0.61	0.63–0.69	0.78–0.90	1.55–1.75	10.4–12.5	2.09–2.35	.52–.90

Spatial resolution is usually given in a distance measurement. For example, most SPOT sensors have a resolution of 10 m; some have a higher resolution of 2.5 m. The resolution does not mean that an object of that size can be *consistently* detected and identified. Various atmospheric and situational characteristics play into this. You might think of this as simply the measure of the side of one of the raster cells detected by the remote sensing technology.

Temporal resolution depends greatly on the spatial resolution of the sensing technology. High spatial resolutions will record a great amount of data for a small area, requiring much longer to return to a place than low spatial resolution sensors. For example, Landsat with 30-m spatial resolution revisits a place only once every 16 days. The Advanced Very High-Resolution Radiometer (AVHRR) has a spatial resolution of 1.1 km and revisits a place once every day.

Spectral resolution is an important characteristic. A coarse spectral reso-

Satellite Image
(30 meter spatial resolution)　　　Arial Photograph
(1 meter spatial resolution)

FIGURE 8.6. Comparison of spatial resolutions.
From *http://www.csc.noaa.gov/products/sccoasts/html/rsdetail.htm*

lution inhibits the ability to detect certain wavelengths and to distinguish features. A finer spectral resolution sensor can be used to represent different features based on the distinct wavelength patterns detected by the sensor.

Classification

Considering these three types of resolutions and other sensor and environmental characteristics, an operator can make a choice about how to classify the pixels from a scene using either supervised or unsupervised classification. *Supervised classification* means that the operator participates in an interactive process that assigns pixels to categories. *Unsupervised classification* occurs automatically without instructions from the operator.

Types of Sensors

This discussion of principles focuses on satellite-based remote sensing technology. This is only part of the available remote sensing technologies. The same technologies used for satellites, or adaptations thereof, are often used for remote sensing technologies used by airplanes, helicopters, and in some case hand-held formats.

Photography

Photography is the most common remote sensing technology. In fact, some of the first military remote sensing satellites used cameras with film in the 1960s. The film was dropped out of the satellite in a special heat-resistance reentry container with a parachute and picked up in the air by an airplane. Satellites still use cameras, but most of the images are now captured and stored digitally. Satellite sensor technologies using photography are "panchromatic" or sensitive to the full visible spectrum. The potential resolutions of photographic images are very high, but may be limited by data acquisition costs. Many governments and companies use aerial photography as a means of data collection. Using ground reference points and calculations to remove subtle changes in the airplane's movements, two aerial photographs taken simultaneously can be used to make a stereoscopic image. They are a very useful type of remote sensing because when viewed with some additional equipment like a stereoscope, it is possible for most people to distinguish heights and elevation changes. A single photographic image that also has the effects of elevation change removed (called planimetric) is called an "orthophoto" and is georeferenced to a coordinate system.

Infrared

Usually when we refer to photographic remote sensing we mean recording EMR in the visible wavelength spectrum, but this can be broadened to include infrared. This can be done with the chemical applied to photographic film (called an "emulsion") or by using digital devices built and calibrated to detect this EMR spectrum.

Multispectrum

The data collected and images made with Landsat, SPOT, and similar sensing technologies are known as multispectrum because of they include different bands. The variability of multispectrum remote sensors opens up a vast number of application possibilities.

Hyperspectral

This type of sensor technology collects more than 16 bands simultaneously. For example, the Hyperion satellite collects 220 bands from blue to shortwave infrared in equal steps (from 0.4 to 2.5 µm) with a 30-m spatial resolution. Flying in formation with Landsat 7, images from Hyperion can be used easily with Landsat 7 images and data.

Radar

Radar is an important remote sensing sensor type. Its ability to penetrate through cloud cover and into the ground make it very useful for applications in areas with frequent cloud cover and for geological work.

Laser (LiDAR)

Not used on satellites, but on planes, helicopters, and from the ground, LIght Detection And Ranging (LiDAR) uses a laser to generate light pulses, the same way radar uses radio waves. LiDAR is highly accurate and cost-effective for collecting elevation data. Because of LiDAR's speed, hand-held units have been introduced to quickly scan an area—for example, a crime or accident scene.

Applications

Images acquired by satellites have been used to produce local, regional, national, and global composite multispectral mosaics. They have been used in countless applications including monitoring timber losses in the U.S. Pacific Northwest, establishing urban growth, and measuring forest cover. Remote sensing images have also been used in military operations, to locate mineral deposits, to monitor strip mining, and to assess natural changes due to fires and insect infestations.

Data Collection in General

Thinking about remote sensing in a most general sense, we can easily distinguish types of data collection by the platform and by sensor technology. If the remote sensing is based on satellite images or data, in most cases we are likely to have multispectral, hyperspectral, or radar images or data. If it is airplane-based, then we are more likely to have aerial photography, multispectral, or LiDAR images or data. If it is ground-based, then we are most likely to find photography, multispectral, or LiDAR images and data. These rules of thumb have exceptions, of course, and will change as certain types of sensor technology and remote sensing systems become cheaper. They are simply helpful in seeing the relationship between costs, types of data, and application types. Applications in smaller areas tend to use airplane-based or ground-based sensor technologies; larger areas tend toward satellite-based remote sensing.

Coastal Monitoring

An important application area is coastal monitoring. Because of the key role of dynamic processes in coastal erosion, coastal monitoring applications tend to use remote sensing sources that can repeat their observations often. Aerial and LiDAR photography and data may be suitable for smaller areas if the area is generally cloud-free; multispectral satellite images and data may be useful for larger areas, and radar may be used for large areas, or areas with frequent cloud cover.

Global Change

With an increase in average temperatures worldwide, shrinking glaciers, and shrinking ice packs, the study of changes to glaciers and Arctic and Antarctic ice fields has benefited greatly from the use of remote sensing images and data. The frequency of observations helps scientists keep track of changes to ice fields and even icebergs in the water. Detailed observations, combined with measurements on the ground, help researchers monitor minute

FIGURE 8.7. Multispectral sensors produce data and imagery to help monitor and model complex coastal changes.

From *http://earthasart.gsfc.nasa.gov/images/ netherla_hires.jpg*

FIGURE 8.8. A composite of different multispectral data to produce a "picture-like" image of the world.

From *http://earthobservatory.nasa.gov/Newsroom/BlueMarble/*

changes in ice fields. Made available online to other researchers, these measurements, images, and data have become a crucial part of a key area of global change research.

Urban Dynamics

Because of the frequency of observation, satellite-based remote sensing images and data have proven to be very useful in documenting and assessing the growth of large cities around the world and distinguishing changes and processes. Urban dynamics are complex, but individual changes in a single area can be compared to assess the impacts of various policies and urban planning programs. Data and models developed to understand past growth can be used to make predictions of future growth and to assess alternative policy and planning proposals.

Precision Farming

Detailed remote sensing images and data, from a variety of platforms, are used by farmers to reduce the use of and become more efficient in the appli-

FIGURE 8.9. Aerial imagery (here from a digitized aerial photograph) can show a great amount of detail.

cation of fertilizers and pesticides. Agricultural factors including plant health, plant cover, and soil moisture can be monitored with remote sensing data (see Plate 5). By combining the remote sensing images and data from different sources, deficiencies of one remote sensing system can be made up. For instance, Landsat provides multispectral data on average only once every 16 days for any place in the continental United States and is impaired by cloud coverage, even partially cloudy weather. By using radar data, scientists have been able to help farmers keep track of changing soil and plant conditions more frequently, which is especially critical during particular phases of plant growth (e.g., pollination).

Summary

Remote sensing is the collection of data without directly measuring the object. It relies on the reflectance of natural or emitted electromagnetic radiation (EMR). It has become an important and, in some applications, key means of data collection. The many types of remote sensors can be basically distinguished into two groups. Passive sensors rely on natural EMR; active sensors require an additional source of EMR. Remote sensing involves the complicated calibration of spectral signatures indicative of things or events with various characteristics and capabilities of sensors. Most sensors distinguish the EMR they detect as bands, which refers to specific ranges of EMR a sensor detects. Sensors also distinguish between spatial, temporal, and spectral resolutions. The many applications using remote sensing keep growing.

Increasingly, remote sensing has been making inroads into traditional surveying domains.

Review Questions

1. What does the term *LiDAR* stand for?
2. What does the term *panchromatic* mean?
3. Data from remote sensing is a powerful tool for analysis. What has prevented it from being more widely taken advantage of?
4. What is the oldest commercial satellite system that is still in use?
5. What is the size/scale of geographic area most well-suited to being studied using remote sensing data? Small, medium, or large?
6. What remote sensing technology is being used in modern surveying?
7. What were some early applications for radar-based remote sensing?
8. What is the highest resolution of panchromatic remote sensing data now available?
9. When was remote sensing first used?
10. How is remote sensing data usually stored?

Answers

1. What does the term *LiDAR* stand for?

   LiDAR stands for Light Detection and Ranging.

2. What does the term *panchromatic* mean?

   Panchromatic is a descriptive for all wavelengths of the visible spectrum.

3. Data from remote sensing is a powerful tool for analysis. What has prevented it from being more widely taken advantage of?

   Remote sensing data is costly to produce.

4. What is the oldest commercial satellite system that is still in use?

   The oldest commercial satellite system is called "Landsat." The first satellite of this system was launched in July 1972.

5. What is the size/scale of geographic area most well-suited to being studied using remote sensing data? Small, medium, or large?

   Generally, remote sensing data is most useful for studying large areas.

6. What remote sensing technology is being used in modern surveying?

   Surveying now uses laser sensors for accurately measuring distances.

7. What were some early applications for radar-based remote sensing?

   Some of the first applications for radar-based remote sensing were climate analysis, iceberg detection, and geology.

8. What is the highest resolution panchromatic remote sensing now available?

The highest generally available resolution is less than 1 m for panchromatic remote sensing data.

9. When was remote sensing first used?

Most people consider the use of the telescope in the 17th century to be the first use of remote sensing.

10. How is remote sensing data usually stored?

Remote sensing data is usually stored on computer using a raster format.

Chapter Readings

Conway, E. D. (1997). *An Introduction to Satellite Image Interpretation*. Baltimore: Johns Hopkins University Press.

Gibson, P. J. (2000). *Introductory Remote Sensing: Principles and Concepts*. New York: Routledge.

Lillesand, T. M., R. W. Kiefer, & J. W. Chipman. (2004). *Remote Sensing and Image Interpretation* (5th ed.). New York: Wiley.

Sabins, F. F. (1997). *Remote Sensing: Principles and Interpretation* (3rd ed.). New York: Freeman.

Web Resources

One of the most consumer-friendly remote sensing-based web applications (registration required for full access) is available online at *www.keyhole.com/*

For information about LandSat 7, see the website *www.keyhole.com/*

NASA provides many fascinating images at its website at *http://visibleearth.nasa.gov/*

Documentation of wetland destruction using animations is available online at *http://svs.gsfc.nasa.gov/vis/a000000/a002200/a002210/index.html*

For in-depth discussion of everything related to remote sensing, with an emphasis on Landsat, but covering other sensor technologies in great detail, see *http://rst.gsfc.nasa.gov/*

For information about SPOT satellites, see *www.spot.com/html/SICORP/_401_.php*

Another source for information about Landsat satellites is available online at *http://landsat.gsfc.nasa.gov/*

For an excellent interactive tutorial about various aspects of remote sensing, see *http://satftp.soest.hawaii.edu/space/hawaii/*

A tutorial introduction to LiDAR is available online at *www.ghcc.msfc.nasa.gov/sparcle/sparcle_tutorial.html*

Download Landsat data online at *www.earthexplorer.usgs.gov*

Exercises

1. Uses of Remote Sensing

Based on the discussion in the textbook and lecture presentation, determine with your neighbor three remote sensing applications and the data required for each. What is the spectral, the temporal, and the spatial resolution that each application requires?

2. Remote Sensing Laboratory Exercise

Objectives

To better understand the use and types of remote sensing.

Overview

For your first employment following completion of your undergraduate degree you can choose between helping a company set up a remote sensing service for urban areas wanting quick information about changes in their areas and helping its sister company provide the same service for national parks in the United States. Choose one of the positions and then to make an assessment of the remote sensing data that is available for that service.

You are one of many people preparing a description of a new service. The management will review the descriptions and decide which services it will develop further.

Instructions

This is an important first step for either service and you have to be sure that your assessment is well documented. Your assignment is to prepare a one-page (single-spaced) assessment of the type of service you propose the company will provide using this set of topics as an outline.

1. Describe the service, including the remote sensing systems.
2. Explain the service in terms of the problem or issue you find requires this service.
3. Identify the data requirements (resolution and frequency).
4. Discuss the role of resolution (spectral, spatial, and temporal) in your study.
5. Identify any additional data the service will need.
6. Explain how this service can be developed into other services.

Chapter 9

Positions, Networks, Fields, and Transformations

GI can be changed in many ways, whereas maps are usually very difficult to change—for example, copying a small map showing major cities of North America on a single sheet of office paper to a wall-size poster. Certainly, GI, like maps, is a form of geographic representation based on measurements, observations, and relationships (see Chapter 2). But unlike maps, GI has not been altered through a cartographic representation. Further, and most importantly, GI can be transformed in various ways. Conventions usually guide these transformations in both overt and subtle ways (see Chapter 11).

We can distinguish between three GI representation types: positions, networks, and fields. The key issues for geographic representation and cartographic representation related to the different GI types and transformations between the GI types are the focus of this chapter. Things and events can be represented as all of these types. Each offers different possibilities for recording measurements, observations, and relationships. As a position, a thing or event is represented as a discrete record of location and properties. Not only is the thing or event fixed in space with a recording of its position in a coordinate system, its properties are also recorded based on observations and measurements made at or of that position. Things and events represented as networks are recorded using an ordered arrangement of connecting points, called nodes (usually), lines, and sometimes areas. The things and events must be recorded in association with one of these geometrical network elements. Fields are used for nondiscrete things and events, which include anything that can be observed or sensed, but usually does not have clearly identifiable limits—for example, ozone, CO_2, or soil pH.

Each GI representation type is recorded using a database (see Chapter 6) or computer-based storage. This storage provides various ways to organize and index the GI on the computer, which are usually determined by the soft-

| Positions | Network | Fields |

FIGURE 9.1. Examples for each type of geographic information.

ware. Storage of positions is the most common, as locations usually best correspond to the cartographic representation of a map, but certain fields are more likely to use networks (transportation scheduling and routing) or fields (environmental modeling). The storage of GI usually approximates GI representation types, but when not, it is often possible to transform GI between various representation types.

Its ability to transform GI is the underlying reason why GIS has become such a worldwide success for so many human endeavors. It is always possible to transform one representation type to another using a GIS. For example, cities shown as points in a position representation (Figure 9.1) can be transformed into areas in a field representation. The transformation into a field can also take into account that cities have fuzzy boundaries, not the sharp edges of a point.

The type of GI representation has consequences for quality and accuracy. Transforming the GI representation of cities as points into areas in a field may make it possible to show how cities diffuse into the surrounding area, but cities, even in a fuzzy form, are never perfectly round like the circle that represents them at small scales. Generally, we can distinguish between intra- and interunit qualities in relationship to the geographic information type. Intraquality describes how well the differences between properties of units are represented—for example, population of cities in the categories less than and greater than 100,000 or with the exact count of the population. Interunit quality refers to the reliability that things and events are accurately represented—for example, the extent of a forest or a marsh. The boundaries of a wetland or city created at large scale will be much more accurate than a small-scale state map showing the location of wetlands and cities.

GI Representation Types

The three types of GI representation refer to concepts used by most GIS to represent things and events. Each representation type uses specific storage and indexing formats for recording the GI representation with information-processing technology. This section introduces each representation type, discusses how it used to represent things and events, and explains how, in very general terms, it is stored in a GIS. This section also introduces topology, a foundation for vector GIS.

Position-Based Geographic Representation

Most GI is recorded using a position-based representation as points, lines, or areas (also known as polygons). This type of GI representation corresponds to the geometric primitives used to draw two-dimensional map elements. It is a handy and convenient way to create GI based on existing maps and for people used to working with maps. It is also very useful for many types of analysis (see Chapters 13–15). Of course, it can be transformed to other GI representation types.

Positional GI representations are usually two-dimensional and static. Events can only be shown in terms of positions and characteristics at a certain point in time. Measured properties are (1) either recorded as attributes of a spatial object, (2) are defined by the extent of the property, or (3) are associated with the measured properties of a predefined area (raster). Relationships are either defined by associations between attributes or relationships that can be established and analyzed by transformations. The two most common storage techniques for this type of representation are vector and raster (see Chapter 2).

Animation can be used to show events with position-based GI representations, but it is always based on a series of static geographic representations. Animations that show a series of images, just as frames in a comic, are relatively easy to create and show. However, they may be based on the interpolation of specific changes rather than measurements, which lessen their accuracy.

Vector GI is stored in a variety of ways. The most common format has been what people refer to as the "georelational model." This model is being

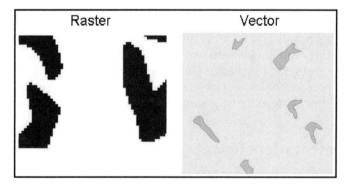

FIGURE 9.2. Examples of raster and vector geographic information representation types.

increasingly replaced by proprietary database storage formats. Although the use of databases is expensive and usually requires specialized organization of the GI and work, they are much quicker than the georelational model storage. However, because of its additional complexity, the traditional georelational model should remain a commonplace fixture of GIS for some time.

The georelational model relies on topology. Topology not only provides a way to reduce the storage requirements for GI, it also provides a means to speed up many processes and check for errors (see Chapter 7).

The georelational model consists of three main components connected topologically. All three components are present and are linked to each other. The first component is a table with a list of polygons (or areas). It records the internal number of a polygon and the chains in the order that make up the polygon's boundary. The second component is the table with a list of chains (also called "lines" or "arcs"). Each chain entry consists of information about the polygons to either side of the chain and the start and end node of the chain. The start and end node define the direction of the chain and which polygons are left and right. The third component of the georelational model is a table of nodes. This table consists of the node identifier and the x and y coordinates of each node.

Additions to the three components of the georelational model can be made to improve the geographic representation and the cartographic representation, especially the addition of additional points used to define the precise shape of a chain and indexes to speed up queries and the drawing time.

Raster GI representation relies on various types of encoding to reduce the amount of storage required by a computer. If each raster or pixel cell is

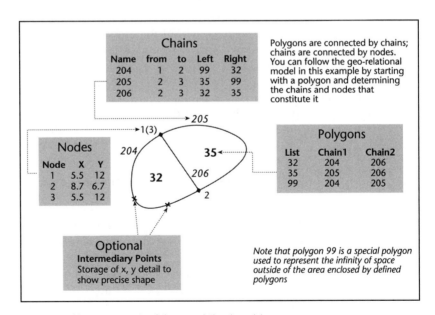

FIGURE 9.3. Key components of the georelational model.

stored individually, the files become very large. A simple way to reduce the required storage (and one of the oldest) is to process each row of the raster data set from left to right, recording only when the attribute value changes and the number of cells following the change to the right. For example, if a row is 100 cells long and cell 1–20 has the value 156, cells 21–78 have the attribute value 123, and cells 79–100 have the attribute value 156 again, the run-length encoded (RLE) raster storage would only store 156:20; 123:59; and 156:21. Other systems are more complicated, but even more efficient. One of the most interesting storage formats is the quad-tree format which works like the RLE approach, but puts areas into a hierarchy of cell value. For example, an agricultural raster data set representing types of crops could distinguish crops at the highest level by the genus, at the next level down in the quad-tree hierarchy it could show the Linnean classification family, and at the third level of the quad-tree it could show individual species. The quad-tree is very efficient and very fast, but changes to the hierarchy can be very complicated and require a great amount of processing.

Network-Based Geographic Representation

The network geographic representation type is usually considered to be a subtype of the position-based geographic information type, but is distinct because of its special properties for representing topological relationships.

The network geographic representation type uses nodes and links, which correspond to nodes and chains in the vector position-based geographic representation type. The distinction is that nodes in the network store information about possible connections (e.g., possible turns at an intersection) and links store the information about how nodes are topologically connected (e.g., Chicago is connected to St. Louis by Interstate 55). Topological information is extremely helpful for vector-based network GI.

Nodes can be added with coordinates from a coordinate system and with additional points with coordinates to define the shape of the networks, for example, situating Chicago and St. Lous on the map in a geographically correct arrangement. However, many networks are represented without this location information, allowing the map to be very simple and easily read (e.g., public transportation maps). (See Plate 6, the London Underground Map.)

Field-Based Geographic Representation

For the representation of nondiscrete, mainly environmental, properties including soil moisture, soil pH, or the distribution of airborne particles and substances including ozone, dust, or pollen, fields are the ideal GI representation type.

Conceptually, fields are nondiscrete, meaning no precise and accurate boundaries can be made between soil pH 6.7 and 6.8, and the properties of a field can be modeled using geostatistical techniques that take these relationships into account, but the storage of the GI representation type usually uses raster data structures. This should always be considered when working with field data. It is easy, but wrong, to interpret raster cell boundaries as the sharp boundaries between different attribute values, when, in fact, the geographic things and events represented by a field are nondiscrete.

A triangular irregular network (TIN) is a specific format for the representation of fields that relies on a network of lines connecting sampled points with known values. The connections form a Delauney triangulation, which means that each point is connected to only two other points to create triangular faces. This type of GI representation is most commonly used for the visualization of elevation data, but can be used for any data that is collected using irregular samples in an area. Dynamic versions of TIN make it possible to rapidly change the TIN. The changes can be so rapid that dynamic TIN holds potential to help train people for complex navigation situations.

Transformations

Even if GI is represented as a field, it may not originate with data collected for every point in the area of the field. Since this detailed data collection would be practically impossible, most field data is usually the result of transforming position-based GI observations and measurements. For instance, a property of soil, pH, shown as a nondiscrete field for an area, may be based on an interpolation of soil samples collected at various points. The soil pH data could be transformed back into a position-based GI representation as contours that show where soil pH changes (e.g., a contour for every 0.5 change in soil pH). Transformations can be applied to any representation of GI. GI can be transformed to different types—for example, positions to fields, or networks to positions, or from one position-based GI representation to another (e.g., points to lines).

FIGURE 9.4. Two examples of field GI. On the left a DEM, on the right GIRAS land use.

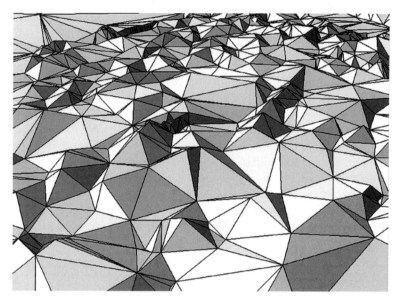

FIGURE 9.5. Example of a TIN data structure. Each triangle is a facet of a hill slope representing a change in elevation, orientation, or the relationship between these two characteristics.

The transformation concept goes back to Tobler's development and application of the mathematical transformation concept to cartography. For Tobler, the map is more than a representation; it is a device for storing information. Tobler worked on mathematical techniques and analytical methods to transform maps into forms of information that can be changed further. Thanks to Tobler's conceptual work, we regard GI not just as data, but as data with meaning, which can be transformed and combined with other GI to create new forms of GI. With the transformation concept comes an understanding of GI as sets of associations with particular representations that can be converted to create other sets of associations.

WHAT ARE TRANSFORMATIONS?

Transformations are operations on GI that change the information content by geometrically manipulating GI and changing it into other GI representation types. For example, a buffer operation can transform a point that represents a well into a polygon that represents the zone around the well. This zone can be represented as positional or field GI, depending on the operation chosen. The zone can be transformed into the other GI representation types. Transformations of GI can also change attributes. An example of an attribute change is converting temperature recorded in degrees Celsius to degrees Fahrenheit. In both cases, the key change involves transforming the GI representation. What information is measured for a point, such as a well, is only of limited validity for an area, such as a theoretical plume extent. A transformation can produce new GI based on calculations that show a relation, as in the example of a buffer.

Examples

The two most fundamental GIS operations, buffers and overlays, are examples of GI representation transformations. Buffers transform position-based GI into other types of position-based GI or fields. Overlays transform two position-based GI data sets into one. What these operations involve and how they transform demonstrates the key role of transformations for GI and its much greater usefulness compared to maps.

BUFFER TRANSFORMATIONS

A buffer transformation is the simplest transformation to grasp, but its operation can actually be quite complex. Practically, based on the position of one or more GI objects, it determines the zone around the objects using one or more distances. Figure 9.6 (left) shows a simple 100-foot buffer around a well. But what do the 100 feet (about 30 m) represent? They may simply be the regulatory zone where no animal waste disposal is allowed. But it could be based on more complex geographical relationships. Maybe the 100 feet corresponds to the well recharge zone calculated using a hydrological model that considers both the soil type and geology. The areas of buffers usually are used to show a geographical relationship. Based on an understanding of the relationship, distances are used to show the extent of the relationship. This technique is used to indicate area affected by vehicle or airplane traffic. Complex models may only use buffers to represent the results of calculations that work with fields and model things and events in

FIGURE 9.6. A 100-foot buffer around a point representing a well produces a vector area or field (left); Buffered zone of land use around the Cincinnati/Northern Kentucky Airport (right).

terms of relationship vectors. This simple operation is a very powerful transformation. In all cases, obviously the accuracy and quality of a buffer depend on the underlying model and explicit (or implicit) assumptions.

OVERLAY TRANSFORMATIONS

GIS overlay is, depending on who you speak with, the first or second most important operation for GIS. Either way, it is without doubt one of the most significant operations. It is also one of the primary transformations, but the transformations performed by an overlay depend on the type of GI representation.

Positional GI combines the geometries (points, lines, or areas) of two data sets based on a common coordinate system. The geometrical transformation is only the geometric process of determining the intersections between objects from each data set and the assembly of new objects that correspond to the original objects. Attributes from the original objects are assigned to the new objects based on the location of the original objects. The

FIGURE 9.7. Overlay transforms by combining two (or more) data sets based on the location of features in a coordinate system.

attribute transformation begins only here. Various operations, logical and mathematical, are used to transform attributes and relate them—for example, evaluating soil type and soil moisture to determine crop suitability. Raster GI performs these attribute transformation as the overlay transformation, assuming both raster data sets use the same raster size and origin point (otherwise some complex geometric transformations must first take place). Chapter 14 covers these issues and the overlay operation in more detail.

Summary

This chapter examined GI representation types and transformations. GI representation types are the formats available for GI: positions, networks, and fields. Positions and networks rely on vector data formats; fields rely on raster data formats. Positional GI is stored in a GIS as points, lines, or areas (also known as polygons), most often following the georelational model that uses topology. Networks also use these data formats, but areas are of very limited use in a network. Points, called nodes in networks, are much more important.

Transformations are operations on GI representation types that change the information content. A buffer transforms a point through a distance measure into an impacted area.

In-Depth Some Applications and Geographic Information Representation Types

Application	GI Representation Type	Data Types
Water pollution	Raster	Water characteristics, models of pollutant diffusion
Vehicle routing	Network	Roads and highways
Biotope conservation	Position	Location and types of biotopes, landuse and landcover

---

In-Depth Topology

A modern branch of mathematics with great impacts in many fields, topology has been an important influence on the development of GI. Topology was introduced by one of history's greatest mathematicians, Leonard Euler, in 1736 when he published a paper on how to solve a puzzle that had perplexed residents of Königsberg (now Kaliningrad). The puzzle sought a solution about how to cross seven bridges that connected two islands in the middle of the city without crossing any bridge twice.

Euler's solution was to abstract the problem into a set of relationships between vertices (also called nodes), edges, and faces. This is called a graph. Euler established that a graph has a path traversing each edge exactly once if exactly two vertices link an odd number of edges. Since this isn't the case in Königsberg there isn't a route that crosses each bridge once and only once.

Figure showing Euler's seven bridges of Königsberg problem.

The mathematics of this relationship are simple. To determine if there is single relationship, count the number of vertices connecting three edges. If the number of vertices is two, then there is a single way around. Otherwise, at least one vertice must be crossed twice.

Euler contributed an immense body of work, over 775 papers, half of which were written after he went blind at the age of 59. The Königsberg problem is related to Euler's polyhedral formula, which is the basis for determining topology in a GIS:

$$v - f + e = 2$$

v stands for the number of vertices, f for the number of faces, and e for the number of edges. Regardless of the type of polygon, this number will always be two.

Topology was extended by numerous mathematicians in the late 19th century, and although most people learn little about it, it has been immensely significant for many technological developments.

Topology focuses on connectivity. In regards to GIS, topology is important for three reasons. First, it can be calculated to determine if all polygons are closed, lines connected by nodes, and nodes connected to lines. This allows for the determination of errors in digitized or scanned vector data. Second, it can be used in network GI to determine network routing. Finally, because it allows that the same line (edge) is used for neighboring polygons, the number of lines stored in a GIS can be greatly reduced.

Review Questions

1. What sets GI apart from maps in terms of discrete and nondiscrete information?
2. Why are multiple types of data structures needed?
3. What is Tobler's transformational concept?

4. What is the main difference between discrete and nondiscrete GI?
5. What is the main difference between topological and nontopological vector data?
6. What is a quad-tree?
7. What is a triangular irregular network (TIN)?
8. How can the GI storage format impact GI representation?
9. How does a buffer operation transform a geographic representation?
10. Why can't maps be transformed?

Answers

1. What sets GI apart from maps in terms of discrete and nondiscrete information?

   GI offers multiple ways to store and transform data that can be used to make meaningful representations of things and events as GI. Maps can show both discrete and nondiscrete information, but the information cannot be transformed.

2. Why are multiple types of data structures needed?

   Different types of data structures make it possible to adequately geographically and cartographically represent observations of things and events.

3. What is Tobler's transformational concept?

   Tobler's transformational concept is the development and application of the mathematical transformation concept to cartography. With this concept comes an understanding of GI as sets of associations with particular representations that can be converted to create other sets of associations.

4. What is the main difference between discrete and nondiscrete GI?

   Discrete GI shows things with fixed boundaries; nondiscrete GI shows processes or states of processes.

5. What is the main difference between topological and nontopological vector data?

   Topological vector data has a set of relationships between nodes and links; nontopological vector data maintains only start, possibly intermediate, and end points.

6. What is a quad-tree?

   A quad-tree is a data structure for the efficient storing of raster data following a hierarchy based on areas of contiguous attribute values.

7. What is a triangular irregular network (TIN)?

   A Tin is a data structure for storing GI based on distance relationships and single values; it is most widely used for storing and modeling elevation data.

8. How can the geographic data structure impact GI representation?

   It allows certain attributes and relationships to be better stored than others; transformations make it possible to convert GI to other formats that may resolve the limitations with one particular type of data structure.

9. How does a buffer operation transform a geographic representation?

Based on existing geometry (point, line, area) and attribute value(s), it creates a new area that represents a new geographic representation with a new thing or event.

10. Why can't maps be transformed?

Maps cannot be transformed because of the cartographic representation and recording in the fixed media of a map. Maps cannot be directly transformed into other representations. Information collected from maps through digitization can, however, be transformed.

Chapter Readings

The second edition of this text contains a wealth of new and additional information, but the first edition is still a classic. See

Burrough, P. A. (1987). *Principles of Geographical Information Systems for Land Resource Assessment.* Oxford, UK: Oxford University Press.

From the computer science perspective, this is a key book documenting the development of GIS:

Worboys, M. F. (1995). *GIS: A Computing Perspective.* London: Taylor & Francis.

This book presents the use of databases for representing GI:

Rigaux, P., M. Scholl, et al. (2002). *Introduction to Spatial Databases: Applications to GIS.* San Francisco: Morgan Kaufmann.

Web Resources

For an introduction to some of the fundamental GI representation issues, see the Wikipedia entry online at *http://en.wikipedia.org/wiki/Geographic_information_system*

For a paper that discusses some of the limitations of the widely used types of GI representation, see the website *www.ucgis.org/priorities/research/research_white/1998%20Papers/extensions.html*

A basic GIS tutorial can be found online at *www.gisdevelopment.net/tutorials/*

Some examples of how animations help visualize the temporal aspects of geographic things and events in current GIS can be found at the website *www.farmresearch.com/gis/gallery/animations.asp*

Exercises

1. Euler's Seven Bridges Problem

Description

Topology is a field of mathematics where distance is not relevant. In this exercise, you will examine some of the basic concepts of topology.

Exercise Instructions

On this rough map illustrating the seven bridges of Königsberg problem that motivated the mathematician Leonard Euler to develop topology, try to draw with your pencil in one continuous line a way to walk around the city crossing each bridge only once.

Questions

Why do you think this is so difficult?
Is it possible?

A second case: assume a flood washes out one of the bridges in Königsberg, leaving six. Draw a route around the city now using one continuous pencil line.

Questions

Does it matter which bridge you take away?
What if you add bridges?

2. EXTENDED EXERCISE: Networks, Topologies, and Route

Overview

In this exercise, you will use GPS equipment to determine the locations of several key points near campus, determine the time to walk between locations, and prepare a network graph where distance corresponds to time.

Objectives

Learn how to apply topography in geography.

Exercise Steps and Questions

1. Configure the Equipment

   Make sure the GPS is working properly. Please check the battery status on the main menu (you get there by pressing the main menu button twice). If the power indicator is significantly below 25%, please see the instructor to get new batteries.

2. Collect Location and Travel-Time Data

   In this step, you will need to collect locational data for each point and the time it takes to walk between each location. *You should plan on taking 1 hour to collect this data.* On the table on the next page, first write down the names of seven *places* (the first place should be in front of the main entrance to the building where class normally meets) you will collect location data for. Go outside the main entrance and wait until you have excellent GPS satellite reception (your accuracy should be less than 30 ft). Write down the coordinates displayed on the GPS receiver, the departure time, and then start to walk at a comfortable pace to your second location. When you get there write down the arrival time and location information. When done writing this information, write down the departure time and proceed to the next point.

   Please note:

   - Each point should be at least 200 m from any other point—further is even better.
   - Each connection between points should only be recorded once.
   - If you walk in the order of your locations, your arrival and departure times are always related. However, if you change the order, you will need to make a note of that on the worksheet.

	Description	Easting	Northing	Elevation	Arrival Time	Departure Time	Difference
1							
2							
3							
4							
5							
6							
7							

Comments/Observations:

3. Make a Map of the Locations

   On a separate sheet of paper, prepare a drawing showing the geographic locations of the data you observed above and the routes you traveled to each point. Make sure to indicate scale and include a legend that explains the symbols you used.

4. Make a Network Graph of Locations and Travel Times

   In this step, you will create a schematic drawing of the seven locations from step 2 on a new sheet of paper. You should arrange the locations in a fashion similar to the graphs showing topology.

   Because this network shows travel time, you want to show the distance between locations as a scale equivalent. Each location should be labeled. Make sure to determine the appropriate scale before drawing the graph: assuming the shortest travel time between locations was 5 minutes and the average travel time was 12 minutes, you want to have a scale that fits all your points on a 8.5″ × 11″ sheet of paper. If the maximum travel time is 30 minutes, a scale of 1″ = 5 minutes will need 6″ on the paper. *You should do this in pencil at first in case you need to make changes.*

5. Draw a Network without Scale

   Based on the network graph from step 4, draw a network graph that is not scaled, but only shows the connectivity between locations. You should still arrange the locations in a fashion similar to the graphs presented in the lecture on topology, but don't scale the distances by time.

6. Evaluate Your Network Graph

   Using Euler's Characteristic,

   $$v - e + f = 2 \quad \text{(evaluate your graph from step 5)}$$

   where v is the number of vertices of the polyhedron, e is the number of edges, and f is the number of faces (remember that vertices are your locations and intersections of edges, edges, the connections between vertices, and faces are the areas bounded by connections). Note: Always add one face for the surrounding area of the network.

   The value should be 2. If it is not, check your graph to make sure you have included all locations and added vertices where paths meet.

Questions

1. Is your graph all on the same elevation? If you need to consider multiple elevations—for example, to show overpasses or bridges—how would that change the connectivity of your graph? What is the term used for graphs that are on the same elevation or level?
2. Copy your scaleless network graph to another sheet of paper and indicate how you would traverse the network in a single trip. If you can't, indicate which node you would have to cross twice.
3. How long will it take you to get around your network once?
4. You have a geographic map, a scaled network graph, and a scaleless network graph for a portion of the campus. What is the map better for and what is the graph better for? What does the use of a scale add or detract? Is it necessary to show network connectivity? Please identify two activities for the each map and each graph.

Part III

Advanced Issues in GI and Cartography

Chapter 10

Cartographic Representation

This chapter covers principles of cartographic representation including scale, generalization, classifications, media formats, and cartographic presentation types. Whether you make GI or maps, cartographic representation is usually a very important concern. The design of maps and visualizations is a critical part of successful geographic communication, which has two components. *Geographic representation* abstracts things and events from the world (see Part I). *Cartographic representation* creates visual techniques and forms for geographic representation. Communication, broadly understood, is how we share a representation with others. Depending on the means and modes of communication, different cartographic representations could be required. A classroom wall map of Europe or North America will show things and events differently than a small map of Europe or North America on a television news broadcast.

Key concerns for cartographic representations center on the design of maps and other material that reliably abstract from and with geographic representations. For this reason, cartographic representations are in many ways inseparable from geographic representations. Good communication often requires thinking of both simultaneously, although the process of creating GI is often separate. In the end, a cartographic representation should maintain the integrity of the GI. Otherwise substantial errors and distortions can occur. This chapter looks pragmatically at the principles of cartographic design as a part of cartographic representation.

Maps and Visualizations

The focus of this chapter requires a distinction between maps and visualizations. Maps are well known to all: they are two-dimensional, or limited three-dimensional, static, and usually printed on paper or some material with many of the same properties of paper. "Visualization" refers to a broader concept, which is mainly connected to images, especially dynamic visualizations, made with computers. Connecting the concept of visualization to the type of media used in communication helps in distinguishing differences at a pragmatic level, but a broader consideration would holistically consider both maps and computer images to be part of systems that also rely on other technologies for visualizing geographic things and events. Following the distinction made here, "visualization" refers to two other important properties for cartographic communication. Display scale, the first property, points to the constraints and capabilities of electronic displays ranging from cellphone displays to LCD projector. The size and resolution of these displays is often limited, necessitating the use of user-controlled zooming functions to change the focus of the display. Second, the tangibility and temporality of visualizations are often fleeting. What is on the display one second may be gone the next, and returning to previous images may be impossible or cumbersome.

Alan MacEachren, a cartographer, developed a conceptual framework for thinking through cartographic representation issues and the important roles of cartographic representation. His work focuses on various forms of map use for cartographic communication or visualization as communication. All map use involves visualization, or cartographic representation, but map use also varies in terms of its relationship to public and private spheres, human interaction with maps, and how the map is used for presentation or discovery. These aspects of cartographic representation should not only contextualize geographic representation, but should also be included among the factors considered when preparing maps or visualizations.

Pragmatic Cartographic Representation Issues

For the pragmatic creation of GI and maps, you need to understand some basic issues and their roles for maps and visualizations.

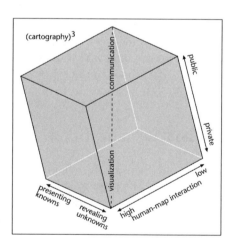

FIGURE 10.1. MacEachren's Visualization Cube. From MacEachren (1994). Reprinted by permission of Alan MacEachren.

In-Depth Suggested Map Elements

Cartographers would likely argue for days and nights about which elements are necessary in a "good" map. For pragmatic purposes, there are six elements that people agree are helpful, even if not always necessary:

- Scale indicators show the relationship between distance measures on the map and the actual ground distance. Scale bars are most common and practical. The representative fraction provides important information for experienced map users.
- Legend explains what the symbols used for a map mean.
- Title provides a simple description of the map, possibly also indicating potential uses and audience of the map.
- Author offers readers the name of the institution, group, or individual responsible for creating the map. It can help point to the acceptance of the map.
- Orientation (North Arrow) helps people orientate themselves when using the map.
- Date indicates when the map was produced and may also suggest when the data was collected.

Explanations and contact information can additionally be included, if they are important for communication, or necessary to ensure that readers can find relevant information for more specific questions.

Note that these six map elements do not necessarily refer to visualizations. Although they can be considered for a visualization, the constraints on a visualizations of size and length of display mean that map elements should only be used when absolutely essential to communication.

Scale

The geographic area shown on a map must be at a scale. The selection of scale and the use of scale have many consequences. The most important thing to remember is that large areas are shown with GI or in maps and visualizations at a small scale, which small areas are shown at a large scale. This may be at first glance counterintuitive; however, scale always states the relation between one unit of distance on the map and the same unit of distance on the ground. "Large scale" refers to a large ratio between map and ground units; "small scale" refers to a small ratio.

Scale for maps and GI, in other words, is usually expressed as a mathematical relationship. However, this is only one of three ways to express scale. Scale can also be expressed in words—for example, "1 inch equals 1 mile" or by using a scale bar.

The representative fraction is the most important way to represent scale. It offers a clear indicator of the relationship between distance measurements on the map and distances on the ground and vice versa. For instance, many topographic maps from around the world are published at a

1:24,000-scale map	1:100,000-scale map	1:250,000-scale map
covers 49 - 79 miles^2	1,568 to 2,240	4,580 to 8,669
(USGS 7.5 minute series)	(USGS 30 x 60 minute series)	(USGS 1 degree by 2 or 3 degree)

FIGURE 10.2. Some map scale examples and sizes covered by USGS map sheets. Images from *http://erg.usgs.gov/isb/pubs/factsheets/fs01502.html*

scale of 1:25,000. The maps at this scale are almost always metric, which means that distance on the map and on the ground is measured using meters or a unit of measurement related to the meter. A word expression of scale of these maps is "one centimeter on the map equals 250 meters on the ground." Another word expression for this scale is "four centimeters equals one kilometer." In fact, for ease of reference, these maps were often called "four centimeter" maps. But the words are still long. Every time the map would be described, people would need to remember the phrase. If people weren't familiar with the phrase, a great deal of explanation would be needed. If we express the scale as 1:25,000, it's much easier. This means that

Representative Fraction	One Inch equals	One centimeter equals
1:24,000	2,000 feet (exact)	240 meters
1:50,000	4,166 feet	500 meters
1:63,360	1 mile (exact)	633.6 meters
1:500,000	8 miles	5 kilometers
1:1,000,000	16 miles	10 kilometers
Scale Bar Example		

Scale Bar Example:
1000   0   1000   2000 Miles

FIGURE 10.3. Some scale representations.

1 mm on a map is equal to 25,000 mm on the ground. There are 10 mm in a centimeter, so 1 mm on a map is also equal to 2,500 cm. With 100 cm in a meter, we can finally convert the relationship of map measurements to ground measurements, or 1 mm equals 25 m. Knowing this relationship, we can easily calculate how long a bridge 4 mm on the map is on the ground. Knowing how many millimeters are in a centimeter, we can also easily calculate how wide a 6-cm field is—or any other distance for that matter.

When using inches, feet, yards, and miles, the mathematics are more complicated, but the principles are the same. The main numerical relationship to know is that the 5,280 feet, or 1,760 yards, of a single mile are equal to 63,360 inches. In other words, at a scale of 1: 63,360, 1 inch on the map is equal to 1 mile on the ground. The calculation of distance measurements from a map or visualization to the equivalent ground distance when using standard or imperial distance units must always first determine the number of feet on the ground in 1-inch distance measured on the map. For example, 1 inch measured on a map in the United States at the scale of 1: 24,000 is equal to 24,000 inches on the ground, or 2,000 feet. If the distance between two intersections on a 1: 24,000 scale map is 6 inches, the distance between them on the ground is 12,000 feet or a little more than 2.25 miles. You should note that this imprecision in reporting distances is just one of the many reasons for using the metric system for distance measurements. The ease of calculations is perhaps the second reason.

Scale can also be expressed by using a bar that indicates a single distance or multiple distances on the ground. A visual comparison offers easy approximation of distances and the use of a ruler to determine the equivalent distances on the map makes it possible to determine the representative fraction and to do similar calculations.

Scale is also a very significant indicator of the detail shown by a map or visualization. Generally, smaller scales show less detail, while larger scales show more detail. Although plenty of cartographers will disagree, many would agree that maps at scales greater than 1: 50,000 scale are called small-scale maps, while maps at scales between 1: 100 and 1: 50,000 are called large-scale maps. There is some disagreement among cartographers about this, but this division provides a meaningful starting point for the discussion.

Finally, different disciplines work with a distinct scale or possibly a set of scales. These scales are usually related to established disciplinary conventions. Because of the cost of GI and the standardization of topographic maps in most European countries, many European planning activities rely on scales of maps created by their national mapping agencies. For smaller areas, they rely on scales standardized for cadastral mapping. This is starting to change as GI becomes more common, but the 1: 25,000 and 1: 50,000 scales are still prevalent.

DETERMINING SCALE BY OTHER MEANS

If you don't know the scale of a map or visualization, you can figure it out by using one of three techniques.

FIGURE 10.4. Calculating map scale using a known distance on a map or visualization.

- Use of known features
- Use of lines of latitude or longitude
- Use of map object comparison

The use of known features uses the ground distances that are clearly known for an object on the map or visualization. Based on the measurement of the distance on the map for the same feature, the map scale can be easily calculated. Because of distortions of paper or minute measurement error, often the results may be slightly in error.

The known distances between lines of latitude or longitude can also be used to determine scale. This approach is especially suitable for small-scale

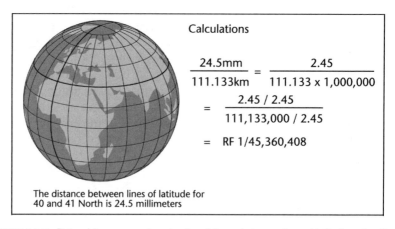

FIGURE 10.5. Determining map scale using the distance between lines of latitude or longitude.

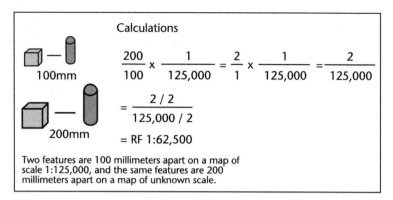

FIGURE 10.6. Map scale can be determined by comparing the distance between two features on two maps, when the scale is known for one map.

maps. The distances between lines of longitude or latitude can be determined from tables published on the Internet.

Comparing distances between the same features on two maps when the scale of one map is known is the third technique. Needless to say, great care has to be taken when making measurements.

Generalization

To cartographically represent things and events, they must be abstracted to fit the target scale or scales and still accurately retain key properties and relationships. Just as scale is an indicator about the amount of detail in a map, it is also an indicator about the amount of *generalization*, the abstraction of features to reduce complexity in maps or geographic information. Generalization serves to assure that cartographic communication works as well as possible at a particular scale or scales. Scale first defines the relationship between an area on the ground and an area on maps or in visualizations; second, scale constrains the possibilities for showing things, events, and relationships.

Generalization is often the most important part of cartographic representation because it assures the role of the map or visualization for the process of cartographic communication. It involves the alteration of GI and symbols. The underlying graphical issue is that with decreasing map scale, graphical elements (points, lines, areas) will refer to larger things and events. A square, 1 mm on each side, can represent a playground 25 by 25 m in size on the ground at a scale of 1:25,000. At a scale of 1:100,000, the same 1 mm square shows a playground 100 by 100 m large.

Two other issues make generalization necessary. Because the clear cartographic representation of things and events can take up disproportionate space at smaller map scales, generalization is necessary to avoid cluttering the map or visualization and to clarify important relationships. Generaliza-

In-Depth Cartographic Representation and Geographic Information Types

Most cartographic representations work with GI stored on a computer. Generalization describes techniques for changing the cartographic presentation. It is a type of transformation, but focuses on graphical transformations (although, if used later as GI, the graphical changes will become part of the GI). The types of GI greatly influence the cartographic representation choices. Ideally, these issues are addressed when first collecting the GI. The desired forms and formats of the cartographic output can be considered and data correspondingly collected.

In many cases, however, the cartographic representation relies on GI collected for other purposes. Vector GI representation types are most common, as they best reflect traditional cartographic practices. Points will be used for the cartographic representation of things and events at small scale; at large scales they may be represented as areas—for example, as towns, buildings, or monuments. Lines of vector and network GI types usually remain lines, but may be strongly simplified at larger scales. This can be a major limitation in representing network relationships. Conversely, raster GI representation types (of both discrete and nondiscrete GI) may need to be aggregated at smaller scales to highlight key properties.

GI types can always be transformed to other types. This often provides an adequate solution, but care needs to be taken in order to assure that the accuracy of the new type will not suggest conclusions that lack support in the original GI. Streets, for example, are often made available as street centerline data. The actual width of the road is not indicated by these lines, nor is the attribute for the street width usually included. A generic width can be assumed based on street type, but clearly buffering the street centerlines by this distance is an estimate at best in most cases and erroneous at worst. A note explaining the process of transformation and pointing out the error should be included and displayed on any maps made with the transformed GI.

---

tion is also needed because of emphasis on certain thing, events, or relationships that are central to the communication role of the map or visualization.

Work on generalization in cartography recognizes a number of different operations. The typologies of generalization operations vary, but certainly distinguish between points, lines, and areas. Other issues reflect institutional, functional, or conventional understanding of the cartographic representation and geographic representation that the generalization should support. To introduce the generalization operations, the five most commonplace operations offer sufficient insight into the significant role of generalization for cartographic representation and communication.

AGGREGATION

Aggregation is used to reduce complexity. It can either be used to merge features that border one another (e.g., land use types) or to group feature that

FIGURE 10.7. Generalization is important to make sure that maps can be correctly used.

From Amair Chaudry, University of Edinburgh, private communication.

are nearby (e.g., buildings belonging to one factory). A potential consequence of aggregation is that detailed characteristics can be lost, important relationships homogenized away, or misleading presentations made.

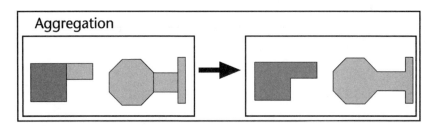

FIGURE 10.8. Illustration of aggregation operation.

DISPLACEMENT

Displacement involves changing the position of points, lines, or areas to either clarify important characteristics, to avoid conflict with other features, or to resolve clutter. It is commonly used for emphasizing the relationship of roads to other features on small-scale transportation maps. Because it results in changes to the positions of things and events, GI collected from small-scale maps needs to be carefully verified against larger scale GI for the same area. The potential distortion of actual geographical relationships should be taken into consideration when working with any map or visualization.

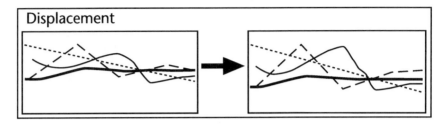

FIGURE 10.9. Illustration of displacement operation.

ENHANCEMENT

Enhancement exaggerates geometrical characteristics of lines or areas to clarify the geographic characteristics or help in inferring geographical relationships from the cartographic presentation. As with the displacement operation, it creates positional changes that distort the relationship between the features and actual things and events on the ground.

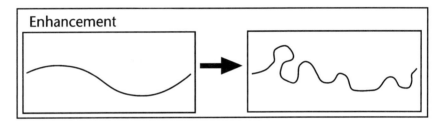

FIGURE 10.10. Illustration of enhancement operation.

SELECTION

Selection is used to retain certain points, lines, or areas for the cartographic representation and to remove others. This mainly aids in dealing with clutter, although this operation is also used to help clarify relationships by removing details. Naturally, this operation can result in significant distortions and possibilities to misinterpret geographic relationships.

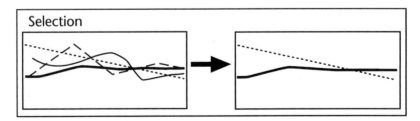

FIGURE 10.11. Illustration of selection operation.

SIMPLIFICATION

Simplification reduces the amount of geometric detail for individual lines or areas. This can be done by arbitrarily removing points that describe the shape of a line or areas, or can be done with several algorithms, or can be done by hand. In each case, varying degrees of changes to the positions result, leading to possible distortion.

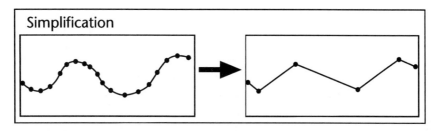

FIGURE 10.12. Illustration of simplification operation.

Classification

The characteristics of things and events stored as GI, or associated with the corresponding GI, need to be clearly organized for communication. Humans can only understand a certain amount of information at once. The rule of thumb for the cartographic representation is to use three to seven categories. Of course, this "rule" is violated plenty of times. It really serves as a guideline that is helpful for thinking about cartographic design. For car navigation or mobile phones, three categories is probably the ideal starting point. A land-use map or study of soils will require many more categories, which does make such maps harder to read, but only then can the map meet the professional or scientific needs for which it was created.

Classification is used for quantitative data, including counts, measurements, and calculations. Ordinal, interval, and ratio data can all be classified. Nominal data can also be classified, but only as individual categories or by clusters—for example, land-use types by generic land use categories, such as urban, mixed, and rural.

Classification is important for cartographic representation. Some people may even claim that its main use in creating choropleth maps makes it at least one of the most important cartographic representation techniques. A choropleth map uses the boundaries of geographic units (e.g., counties, countries, or states) to determine the area represented with a particular shade or color. Since the geographic units are distinct, the cartographic representation makes it easy to compare the characteristics of each unit (see Plate 7).

Choropleth maps and visualizations start with defined geographic units, for which data is either collected, or, if the data comes from other data sets, is associated with the same geographic units. These units may be counties, countries, or states, but they can also be an archaeological site, biotope, or watershed. It is of course possible to aggregate the data from one type of geographic unit to a larger unit (e.g., counties to states), but different areal aggregations might result in very different representations, supporting very different conclusions. When creating a choropleth map, it is very important to state which geographic units were used for collecting the data (see Plate 8).

In urbanized areas and for environmental applications that rely on administrative or arbitrary boundaries, these issues are of special importance. These boundaries frequently can divide communities or congruent zones—for example, a lake in the middle of a desert. While the water is regionally significant, the lake should only have minute influence on the statistics of a largely desert area. However, if most of the lake ends up in a small town, taken by itself, it may look like the town has rich water resources to itself.

TYPES AND APPLICATIONS OF COMMON CLASSIFICATIONS

Equal Interval. The equal interval classification divides the total range (the number between the minimum value and the maximum value) into equal parts. This classification is best used for properties of things and events that have an implicit order—for example, the top 30 stores by sales in the state. This classification is valuable for highlighting such an order, but can mask details and deviations.

Municipality Areal Units

Number of Inhabitants

24	56	23	17	6	45
21	28	17	21	9	35
27	45	19	39	7	49
25	41	11	27	3	41

Number of Dogs

1	5	9	4	35	10
5	21	7	2	3	14
3	15	4	9	10	12
4	18	5	4	19	17

Dogs per Inhabitant

0.04	0.09	0.39	0.24	5.83	0.22
0.24	0.75	0.41	0.10	0.33	0.40
0.11	0.33	0.21	0.23	1.43	0.24
0.16	0.44	0.45	0.15	6.33	0.41

Aggregated to County Areal Units

Number of Inhabitants

267	174	195

Number of Dogs

72	44	120

Dogs per Inhabitant

0.27	0.25	0.62

FIGURE 10.13. Areal aggregation of geographic units can lead to results that mask important geographical characteristics and relationships.

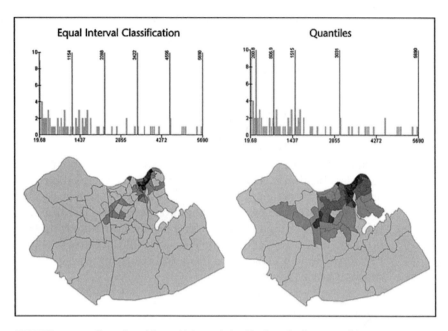

FIGURE 10.14. Examples of four widely used classifications for the same data.

FIGURE 10.14. *(cont.)*

Quantile. The quantile classification determines classes to ensure that each class has approximately the same number of features. This classification produces results that best show the diversity of a geographic property and can aid in determining relationships, or can lead to questions about possible relationships.

Natural Break (Jenks). Also known as the "Jenks classification" after the cartographer who developed it, natural breaks use an algorithm to determine where class boundaries should be placed in the total range. The class boundaries should maximize differences and keep similar clusters together. This classification is very valuable for visualizing GI in an insightful manner.

Standard Deviation. The standard deviation classification is different from the other three. It shows graphically how much a geographic unit's property varies from the mean. This is useful for cartographically representing differences in geographic properties from an average value rather than showing exact values.

Symbolization

Things and events require additional abstraction for certain mapping and visualization purposes. The symbols for elements from the geographic representation can be varied in terms of size, shape, value, texture, orientation, and hue. These symbols are used to represent location, direction, distance, movement, function, process, and correlation in an infinite number of ways. The design of cartographic materials draws on long traditions and scientific study of these symbols and their role in cartographic representation and communication.

Pragmatically, size, shape, and hue are perhaps the most significant. Size is important because humans will instinctively understand that a larger dot on a map of the world means a larger city than a smaller dot or that a large runway symbol indicates a large airport. Size shows quantitative differences. Shape provides clues about qualitative differences. Shape is also important because it gives us an orientation to the actual shape on the ground of the feature and because it helps us to understand geographic relationships. Hue, which refers to what most people refer to commonly as "color," is so significant that it requires its own section.

Color, Otherwise Known as Hue

A key component of most maps is color. Humans have an astonishingly broad perception of color, but a surprising number of people face limits in distinguishing color (color-blind people and others with visual or perceptual impairments) and many people will not agree on the name of specific colors. Color is very important for maps and visualization. A good application of color will greatly help with communication.

The first principle to remember is that the use of color should be dis-

FIGURE 10.15. Size and shape are key visual variable to distinguish quantity and quality in cartographic representation.

crete, never garish, and never use overly bright colors that many people (especially the elderly) have difficulties distinguishing. Generally, all people have great difficulties putting colors into a sequence, making color ill-suited for showing quantitative differences. People distinguish gray tones, or value, as a sequence from light to dark. The second principle is that colors should align themselves with existing conventions. All people associate blue with water, but other colors may be associated with cultural values and, in some countries, laws require the use of certain colors for certain legal documents and maps.

Color is very complex and differences in displays and printing equipment can lead to astonishing differences. Different standardized color models are generally used for web documents, video, computer screens, and print documents. CMYK (cyan, magenta, yellow, and key—black) is the most common specification for print; RGB (red, green, blue) is widely used for screen displays.

Cartographic Representation Types

Media and Formats

Depending on the media and the format of the map or visualization output, cartographic representation must be adapted. Size and format are important, but the media is a key variable that should not be overlooked.

OUTPUT TYPES

The output types of maps and visualization usually range from small newspaper, cellphone, or brochure maps to large poster-size maps. Specialized

---

In-Depth Polishing the Cartographic Image

Cartographers might not agree on many points of cartographic design; however, a few suggestions can be a great help if you are just starting to make maps or lack the time to take a course on cartographic design. Following Mark Monmonier, who has written a series of widely acclaimed books on the uses and misuses of cartography, the following rules should guide attempts to improve maps.

1. Be shrewdly selective. Don't show what you'd rather they not see.
2. Frame strategically. Avoid unfavorable juxtaposition and crop the maps and sketches to forestall fears of illness or diminished property values.
3. Accentuate the positive. Choose favorable data and supportive themes for maps.
4. If caught, have a story ready. Computer errors and a stupid drafting technician's use of the wrong labels are plausible excuses.
5. Minimize the negative. If you can't eliminate them entirely, at least don't emphasize features you'd rather have ignored.
6. Dazzle with judicious detail. After all, a detailed map is a technically accurate map, right? Details are useful distractions.
7. Persuade with pap. Try simplistic maps, or maps that camouflage potentially embarrassing details.
8. Distract with historic maps and aerial photographs.
9. Generalize creatively. Filter or enhance details to prove your point.
10. Enchant with elegance, use lots of tree symbols.

---

maps can be even larger (e.g., billboard advertisements) or smaller (e.g., postage stamps) and on a variety of material, (e.g., stone carving, metal etching). Most maps and visualizations are either on paper (or a paper-like material) or on some form of computer-controlled display device.

For pragmatic reasons, this discussion focuses on cartographic representation issues only for these most common output types. Paper is still the most common media for cartographic representations. Formats range from small grayscale maps published in newspapers, advertisements, and brochures, to high-resolution large posters of photographic reproduction quality. Small maps need to be simple. If they are complicated, they can become muddled and will communicate poorly. If a small map is printed in grayscale it can easily show quantitative differences, but may become too complicated if many symbol shapes are used.

The output types of visualizations range from small cellphone displays to large computer screens and projectors. However, even if the area of the screen or projected surface is large, the resolution is still very low in comparison to that of printed maps. The lower resolution can be compensated for by offering capabilities to zoom and pan in the visualization. This makes it possible to interactively move around a visualization based on a user's interest. While this is an essential feature for small displays—cellphones and similar devices—it is also necessary for larger displays of visualizations because of their lower resolution.

TYPES OF OUTPUT EQUIPMENT

Output equipment for cartographic visualizations can be separated into two groups. Printing equipment produces output on paper or paper-like material. Visualizations are made with displays.

More and more printing equipment has become available to produce color maps, but the costs remain comparatively high. A color ink-jet printer may cost less than $100, but the ink cartridge for 30 full-color maps may cost another $30. Printing costs of $1.00 for a color map is very expensive. For large series of maps, such costs can be prohibitive. Grayscale maps, printed with laser printers, ink-jets, or photocopied, remain common and will likely be widely used far into the future.

Depending on the resolution of the printing equipment, care should be taken with grayscales. If the printing equipment has a very low resolution (e.g., a fax machine), the clarity of the high-resolution grayscale map may degrade into a collection of unsightly ink spots. Issues with printing equipment resolution also play a role in making large-format maps. Because of differences in printing technologies, maps that portray well when previewed on a computer screen may appear much worse when printed on paper. If a substantial map is being produced, it is wise to check the quality of the final map with a sample before commencing a large print-run.

High-end color printing equipment supports resolutions of over 1,000 dots per inch (dpi) and brings great detail to printing. If the color printing

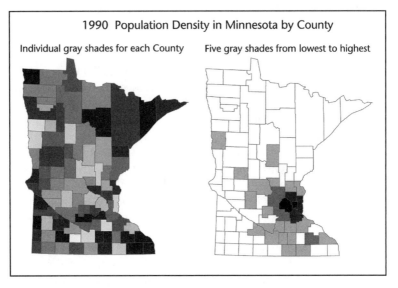

FIGURE 10.16. Ill-advised use of symbols can lead to confusing maps.

FIGURE 10.17. GIS digitization, viewing, and plotting equipment.
Above, from *www.epa.gov/ada/about/thumbnails/gis.htm.*

equipment supports a much lower resolution (e.g., the 300 dpi of most large-format color ink-jets), then the print quality difference will be noticeable, but is usually not enough to distract from engineering or planning applications.

COMMUNICATION GOALS

Clearly the choices of media, format, and output depend on costs and available resources, but the communication goals remain the most important. Whenever possible, the communication goals of a map or visualization should be taken into account at the beginning of working on the geographic representation and cartographic representation.

Depending on media and format, the same communication goals may require different geographic representations. Advertising for a new clothing store in a monthly magazine will need a different map than the store location map on a website or available for cellphone users on request. There is rarely a need to create three different geographic representations to support these different communication goals, but it is beneficial if the geographic representation can consider the goals and create GI to support them all.

The early work on a geographic representation dovetails well with work on a cartographic representation that supports different communication goals. Generalization offers some flexibility for using GI in different ways, but additional effort is usually required to assure that the cartographic representation supports the communication goals. This usually involves the use of different symbology for different output media and formats.

DISTORTIONS

Different output media and formats have different levels of distortion. This can be very important if the maps or visualizations are being used for technical or legal purposes, but the distortions resulting from generalization may be far greater.

Most of all, the process of cartographic representation can produce changes in the position of objects. Road geometry is moved to clearly show the connectivity and to remove conflicts with other feature symbols. These changes may not be indicated and if the altered GI is transferred to others without an indication of the changes, its use for other purposes may lead to erroneous results. It is also possible that attribute changes, usually as a result of generalizing GI, can later lead to grave errors.

Types of Presentations

Creativity knows no bounds and cartographic presentations evidence this creativity in an endless gamut. This section of this chapter presents merely a selection with a pragmatic emphasis on the most common types. Many of these types are combined into hybrids. These types offer a structure to orientate with. Depending on experiences, training, disciplines, and institutions, other organizations will be more sensible.

TOPOGRAPHIC

The "universal" map was developed for military and civilian use in the 16th century by European countries, but in the 19th century it took on a form that made it the most significant kind of mapping for governance, the military, and colonization. The military often commissioned or heavily influenced topographic maps. Significant details and secret information would be edited out of civilian versions of maps. In some countries, distortions were introduced to assure that even generic information could never be used to locate other things or events. Ideally at a scale of 1:25,000, or even 1:10,000 for very detailed maps, the high costs of mapping often led to topographic maps being produced at scales of 1:50,000 or 1:100,000.

CADASTRAL

Cadastral maps show land ownership, rights to land access and use, and obligations. In most countries they are best known for their use in taxation, but this only becomes a significant revenue source in a few countries of the world, especially the United States. The cadastral map's significance comes mainly from its role in governance and in land development and speculation.

FIGURE 10.18. Portion of the 1947–1951 USGS topographic map from Minneapolis, Minnesota.

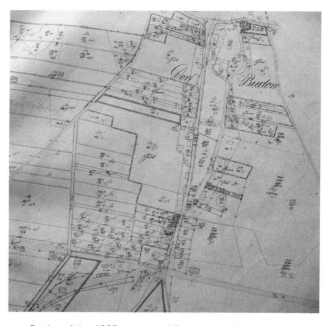

FIGURE 10.19. Portion of the 1865 cadaster of Budow (now Budowo).

A government wishing to pursue developments needs cadastral mapping in order to assure it knows where it can develop and who should be involved. In times of upheaval, the cadastral map has also been an aid for taking land away from groups and individuals and turning it over to others.

THEMATIC

Thematic maps, both qualitative and quantitative, are perhaps the most widely used form of maps. They make it possible to show things and events that people can otherwise not see nor experience—for example, 2003 birth-rates in counties, or population per square mile in all fifty U.S. states (see Plate 9). The major strength of this form of cartographic presentation is that it enables easy comparison. The birthrate, other demographic statistics, or environmental statistics, can be presented in a tabular form, but the use of a cartographic presentation helps with comparisons by making the geographical relationships plain.

CHOROPLETH

A special type of thematic maps, choropleth maps use a graphic variable (usually hue or value) to show quantitative differences. They are easily prone to misunderstanding because they fail to distinguish differences between subareas of the geographic units shown. For example, a map showing birth-rates in the 50 U.S. states fails to distinguish urban and rural differences. Choropleth maps are also erroneously used to show the number of a particular property for an area—for example, population, cars, or voters. However, the area of the geographic unit (e.g., state or county) plays a direct role on how much of the property can be found in an area. No matter how urban an area is, the number of religious buildings stands in relationship to the size of the area. The easiest way to avoid this error is to always use choropleth mapping for properties of a constant unit (e.g., an acre or hectare). This approach was further developed as dasymetric mapping.

CHARTS

Charts often stand alone for naval or aerial navigation, but they can be used in conjunction with maps to provide geographic orientation.

CARTOGRAMS

Cartograms, both contiguous and noncontiguous, show quantitative difference by altering the size of the geographic units according to the relative proportion of the geographic unit's property (see Plate 10). One type of cartogram stretches all boundaries to create an image that roughly corresponds to the original, even if greatly distorted. The second type simply scales the boundaries of each unit according to its rank with other units.

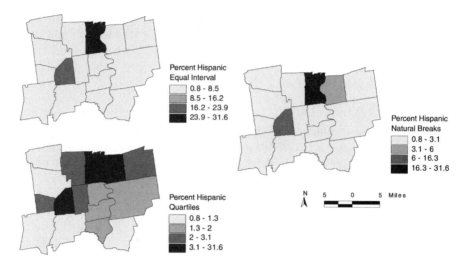

FIGURE 10.20. Examples of choropleth maps of the same data using different class interval methods. From Cromley & McLafferty (2002). Reprinted by permission of Guilford Publications.

DOT-DENSITY MAPS

A dot-density map or visualization is used to show the quantity of a property by placing a dot in a geographic unit for each incident of a property (or multiple of incidents). The dots can either be placed to correspond to the position of the incidents, or randomly placed within each geographic unit. The former provides an excellent understanding of the property's geographic distribution, but must be prepared by hand or with specialized software.

Summary

Principles of cartographic representation include scale, generalization, classifications, media formats, and cartographic presentation types. Geographic representation is how to abstract things and events from the world. Cartographic representation deals with the creation of visual techniques and forms for the geographic representation. Depending on the means and modes of communication, different cartographic representations could be required. A classroom wall map of Europe or North America will show things and events differently than a small map of Europe or North America on a television news broadcast. The design of maps and visualizations is essential to successful geographic communication.

Key concerns for cartographic representation center on the design of maps and other material that reliably abstract from and with geographic representations. Cartographic representations are in many ways inseparable from geographic representation. High-quality communication often con-

In-Depth Dasymetric Mapping

John K. Wright developed this technique for mapping that merges geographic representation and cartographic representation. Recognizing that choropleth maps fail to distinguish between subareas of an aggregation unit (e.g., urban and rural areas in counties), they will show amount of grain grown per hectare for urban areas where no corn or wheat is grown, and show average number of people in a family for rural areas where no people live.

Wright's solution subdivided the aggregation unit into subareas that were assigned weights based on associations between the land-use type of that sub-area and the property to be mapped. The weights add up to 1 for the entire aggregation unit.

The dasymetric technique is a far more accurate way of representing GI than choropleth maps and can be relatively easily done using GIS overlay operations and some calculations. See Plate 11 for an example of the difference between a choropleth and a dasymetric representation.

1990 Town Population
• 1 Dot = 300

5   0   5 Miles

N

FIGURE 10.21. A dot-density map of population density.

From Cromley & McLafferty (2002). Reprinted by permission of Guilford Publications.

ceives of both simultaneously, although the process of creating GI often is separate. In the end, a cartographic representation should maintain the integrity of the GI. Otherwise substantial errors and distortions can occur.

Review Questions

1. What is the ideal number of classes to show on a map?
2. What does the larger size of a symbol generally imply?
3. When is cartographic animation generally useful?
4. Why is it important to know the type of output equipment when considering cartographic representation issues?
5. What makes cartographic symbols iconic?
6. What are five suggested map elements?
7. What should symbolization show?
8. When is an equal interval classification useful?
9. How is a histogram useful?
10. Why is the Jenk's classification widely used?

Answers

1. What is the ideal number of classes to show on a map?

   According to cartographic research, between three and seven classes is ideal.

2. What does the larger size of a symbol generally imply?

   Size should be used only to show ordinal distinctions.

3. When is cartographic animation generally useful?

   Cartographic animation is generally helpful for showing events when only discrete GI is available.

4. Why is it important to know the type of output equipment when considering cartographic representation issues?

   Output equipment can provide specific and constrain choices with significant consequences for cartographic presentation.

5. What makes cartographic symbols iconic?

   The use of commonplace symbols to achieve strong, heterogeneous semiotic references makes cartographic symbols iconic.

6. What are five suggested map elements?

   The five most commonly suggested map elements are scale indicators, title, author, date, and north arrow. Explanation and contact information can additionally be considered.

7. What should symbolization show?

   Symbolization should show enough reference to relevant and meaningful things and events without unnecessary exaggeration and conflict to other symbols. The whole "map" should balance cartographic representation needs with the underlying geographic representation.

8. **When is an equal interval classification useful?**

   The equal interval classification is suited for ordinal, interval, or ratio ranges of attribute values that are evenly distributed.

9. **How is a histogram useful?**

   A histogram is useful for graphically showing the range of attribute values.

10. **Why is the Jenk's classification widely used?**

    Based on statistical techniques to identify clusters in data, the Jenk's classification minimizes variance within clusters. The homogeneous classes usually offer a better cartographic representation with wide-ranging attribute values.

Chapter Readings

Cromley, E. K., & S. L. McLafferty. (2002). *GIS and Public Health*. New York: Guilford.

Kraak, M. J., & F. J. Ormeling. (1996). *Cartography: Visualization of Spatial Data*. Harlow, Essex, UK: Longman.

MacEachren, A. M. (1994). *Some Truth with Maps: A Primer on Symbolization and Design*. Washington, DC: American Association of Geographers.

MacEachren, A. M. (1995). *How Maps Work: Representation, Visualization, Design*. New York: Guilford Press.

McMaster, R., & K. S. Shea. (1992). *Generalization in Digital Cartography*. Washington, DC: American Association of Geographers.

Monmonier, M. (1991). *How to Lie with Maps*. Chicago: University of Chicago Press.

Robinson, A. H. (1984). *Elements of Cartography*. New York: Wiley.

Web Resources

ColorBrewer offers an excellent resource on uses of color in cartography online at *www.personal.psu.edu/faculty/c/a/cab38/ColorBrewerBeta.html*

For information on geovisualization, see *www.geovista.psu.edu/sites/icavis/icavis/ICAvis_overview(1).html*

For practical guidelines for U.S. NPS Mapmaking (Chapter Mapmaking for Parklands), see *www.nps.gov/nero/interpanded/graphics/idt2p3.pdf*

The practical guidelines developed by the US Forest Service are available online at *http://purl.access.gpo.gov/GPO/LPS17532*

Exercise

1. EXTENDED EXERCISE Map Measurements

Overview

In this exercise you will learn techniques for measuring distance and area on maps. An important part of making measurements is assessing their accuracy.

Objectives

Of the different ways to measure distance and area from maps, the most common techniques use specialized instruments (an opisometer or a planimeter) or straight-edges and grids. Accurate measures are important. A variety of processes can negatively impact accuracy. The scale of the map is one such issue. Generalization may also have impacts. Warping and stretching of paper, or imprecision in map production, can also have more important effects. In this exercise you will learn how to make measurements from maps and gain some insight into the results of generalization and other factors on the accuracy of map measurements.

Steps and Questions

For steps 1a and 1b you will need a map sheet at the scale of 1:250,000 and a map sheet at the scale of 1:24,000 that cover the same area. If your map sheet is out of the Public Land Survey (PLS) in the United States, skip the questions that ask for township and range information.

Step 1a—Get a 1:250,000 Topographic Sheet

Find a 1:250,000 topographic sheet for an area that overlaps with the 1:24,000 topographic map sheet you get in step 1b. Also make sure that the 1:250,000 sheet includes all the features you will need to locate in step 1b. Identify a feature (small building, intersection, pond, etc.) on both map sheets that you will use for calculating distances.

What is the name and/or type of your feature? _____

What are its UTM coordinates? _____

With your straightedge, or paper-strip, measure the distance of this feature to three other features:

Distance (in meters) to nearest road intersection _____

Distance (in meters) to highest elevation point _____

Distance (in meters) to nearest township/range corner _____

Step 1b—Get a 1:24,000 Topographic Sheet

You should measure the location of the same feature you worked with first in Step 1a feature:

What is the name and or type of your feature? _____

What are its UTM coordinates? _____

With your straightedge, or paper-strip, measure the distance of this feature to three other features:

Distance (in meters) to nearest road intersection _____

Distance (in meters) to highest elevation point _____

Distance (in meters) to nearest township/range corner _____

1. How large (in page units) is the feature you measured from? How will its size, especially in comparison to the distances you measured, affect the accuracy of your measurements? Will they be more significant than generalization effects?

2. Do you recognize any generalization effects in the smaller scale map? Considering all five types of operations, which ones? Describe or sketch examples.

Step 3—Measuring Area

A number of grids are available for you to use in this step. Take one of the 1:24,000 Scale Grids and one 1:250,000 Scale Half Mile Grid sheet. Each cell of the 1:24,000 Scale Grid is 660 × 660 feet, or 435,600 square feet (equivalent to 10 acres). For our purposes, on the 1:250,000 Scale Half Mile Grid, each cell (approximately 1/8") is equal to the same length and area.

Determine the areas of some topographic features using the two following techniques. If a portion of a feature fully fills a cell, it covers an area of 435,600 square feet. Partial cells can be treated following technique A or B. You will use both techniques so be sure you understand them:

A—For each cell partially occupied by the feature, estimate the proportion of the cell that is occupied. Compute the sum of these values, then add the sum to the area of the completely occupied cells.

B—Count the number of partial cells and multiply them by the area of an individual cell. Then divide by 2. Next add this sum to the area of the completely occupied cells.

To measure area, position the lower-left corner of your area grid at the lower-left edge of the feature you will measure. Determine the areas according to technique A or B.

Determine the size of the following areas in square feet on your 1:24,000 topographic quadrangle. *If you use the "English Area Grid," you should you use the printed conversion factor to determine the area in acres. Otherwise, convert the number of cells using your value from Question 3 or 4.*

The PLSS section your feature is located in:

Technique A _____

Technique B _____

A nearby lake or pond:

Technique A _____

Technique B _____

A city block or town extent:

Technique A _____

Technique B _____

5. Are there any differences between technique A and B? How do you explain them? Which technique do you think is preferable? Make sure to consider the different sizes and shapes of features you measured.

Step 4—Compare Scale affects on area measurements

Find an areal feature (at least 2×2 inches) on your 1:24,000 scale topographic map and measure it using techniques A and B. Make sure you can find it on your 1:250,000 scale map:

Feature name and type _____

Area using:

Technique A _____

Technique B _____

Find the same feature on your 1:24,000 scale map and measure its area using techniques A and B:

Area using:

Technique A _____

Technique B _____

6. Is the feature larger or smaller on the 1:250,000 scale map? What generalization effects do you find? Explain the differences you detect with specific examples (e.g., the road defining the park's northern border is greatly simplified in the 1:250,000 scale map). Make a drawing if it helps.

	Point	Line	Area
Aggregation	X	X	X
Displacement	X	X	X
Enhancement		X	X
Selection	X	X	X
Simplification		X	X
Smoothing		X	X

Chapter 11

Map Cultures, Misuses, and GI

How and what GI and maps represent are important questions related to cultural issues and values; these representations give maps power. GI and maps are not simple devices for communication, but complex symbolic vehicles that play with past and present images, other symbols, and ideological and material concerns. GI and maps can communicate in different ways and at different levels to a number of people. While geographic representation and cartographic representation offer a multitude of choices, many of these choices follow conventions that may be implicit. Often conventions reflect particular cultural values, values that arise in a culture's ways of making sense of the world. In other words, the power of GI and maps is intrinsic to a culture's engagement with the world. GI and maps are not only a means of *understanding* but a means of *influencing* what people know. Some people might like to say that misuses of GI and maps are part of exchanges and conflicts between different cultures. All GI and maps operate and make use of forms of geographic representation, types of cartographic representation, assumptions, and conventions that we can bundle together as GI and map cultures. Knowing the cultures of GI and maps helps develop a better sense of how GI and maps inscribe meaning to geographies, how people manipulate GI and maps, and how the misuses of maps can lead to biases in how we understand the world or parts of it.

The purpose of this chapter is the development of a pragmatic understanding of the culturally influenced problems inherent in geographic representation and cartographic representation in Western civilization. These include accuracy issues discussed in previous chapters. More importantly, these problems touch on important questions of what and how GI and maps represent. Above all, there looms the question of why GI and maps are created. These activities are very costly for all people involved. Why people choose to make GI maps and how and what they choose to include in their geographic representations and cartographic representations are important questions to ask about any GI or map. Other civilizations from East Asia, South Asia, Africa, and the Middle East have contributed greatly to cartography and have their own forms of cartography that we still have much to learn about and from.

Cultures of Maps

Culture is a word with many meanings. To discuss the culture of GI and maps, we can distinguish between "national culture," "indigenous culture," and "disciplinary culture." *National culture* refers to the overarching set of values, beliefs, and implicit understandings held by the majority of people living in a particular nation. *Indigenous culture* corresponds to the beliefs and knowledge of people who lived in an area before it was colonized and their heirs. *Disciplinary culture* indicates a more specific set of values, beliefs, and implicit understandings governing a particular profession or discipline—for example, sanitary engineers and city planners in an urban realm, or conservationists and consulting engineers for environmental issues. Blends of cultures characterize individual perceptions more often than clear-cut divisions between cultures. Many people have multiple professional responsibilities and have different professional and life experiences that impact their own cultural understandings. Regardless of the variety of cultures and difficulties of pinning down the exact influence of any culture, culture remains significant for GI and maps.

Civilizations and Maps

Before looking at more specific influences of culture on GI and maps, we might want to start by taking a step back from specific cultural issues of geographic representation and cartographic representation and exploring the relationships between civilizations and maps. As far as archaeologists can tell, all human cultures have had some form of representing geographic things, events, and the relationships between them. It may take the form of a prehistoric cave drawing showing animals with hunters attacking them, it can take the form of an organized set of sticks bound together, or it can take the form of a story that describes the relationships between a tribe and the universe sketched in the sand, or something else that most people today would also call a map.

EUROPEAN

European cultures and cultures in the Middle East—or plain Western civilization, if we want to avoid dealing with the complex differences and conflicts between the different national cultures of Western Europe and many colonized countries—put maps and their precursors at the heart of how many people came to understand the world. Maps in the past complemented experience in much the same way as today by depicting things and events beyond most people's experience. For example, the Romans relied on maps to show the extent of their empire. The maps became iconic, as an image of the geographic extent of the empire. Before the Romans other cultures also used maps. Sometimes maps took on more of an iconic role, and sometimes maps were simply used for practical tasks.

The Romans developed maps in both their iconic and practical roles further. Iconic maps served, as Denis Cosgrove writes, as a way for individuals to understand their place in an empire and to associate their calling and position in life with both the mundane events of day-to-day life and the divine, symbolized by the emperor for most of Roman history. The cosmology of the Romans involved maps. Practically, Romans not only advanced but solidly established the basic principles of cadastral maps (also see Chapters 5 and 12), which would become the basis for modern arrangements of land ownership and subdivision hundreds of years later. The itineraries used by the Romans became the basis for showing travel times between cities. Maps showing connections based on time are still important for transportation, especially road maps that show the relationships between places.

From the late Middle Ages (around 1400 A.D.) on, maps took on more practical functions for military uses but also for cadastres and increasingly for the visualization of statistical information used for social and environmental decision making. The iconic role of maps remains to this day, as we see in many company logos and advertisements. Maps (and later GI) became part of the increasingly specialized and bureaucratized social activities that required detailed understanding of the world and the ability to share this understanding with other groups and cultures. The military plays a key part in these activities, although we also need to remember that commercial interests have produced or adapted military GI and maps with many significant uses.

INDIGENOUS

Native, or indigenous, cultures around the world have relied on forms of geographic representations that should also be considered as maps. Even though they may be sharply different from what people in Western-influenced cultures have come to understand as maps, they are material devices used to communicate things and events in their geographic relationships. An example of this type of map is the stick charts used by Pacific South Sea islanders as training device and navigation aids for people traveling across the vast expanses of the Pacific Ocean.

Other indigenous cultures had other forms for communicating geographic ideas, often connecting spiritual and physical geographies. The failure to recognize indigenous cultural geographic representations has been an ongoing source of conflict in many areas of the world. Rundstrom, along with many others, points out that the assumption that a European-based representation of the world in a topographic map contains just the "facts" represented by naming can lead to the inclusion of sites and artifacts that hold special spiritual and cultural significance for indigenous groups. As these

FIGURE 11.1. Polynesian stick and shell map.

Department of Anthropology, Smithsonian Institution, catalogue No. 398227, photo by P. E. Hurlbert

sites and artifacts are attractions for people from other cultures, their mapping has lead to people coming and destroying the sites by removing items or disrespecting the indigenous culture's activities.

CULTURAL FORCES WITHIN "DISCIPLINES"

The cultural influences that have the greatest effect on map users probably come from within mapping "disciplines" themselves, even though it may be hard to identify specific values, beliefs, and so on. Surveyors have concepts of geographic and cartographic representation that are different from those held by other geography-oriented professionals. Scale, and with it the minimum resolution of geographic objects, usually will be different, and the selections of things and events for inclusion in the geographic representation will differ significantly. The culture of surveyors focuses on accurately recording the location of specific things—and possibly events related to the purpose of the survey. Generally, surveys are prepared for governments and individuals for the purpose of identifying property lines, and planning for improvements such as buildings and other construction. The culture of planning emphasizes the uses and functions of buildings and other surfaces in an area.

The people who create and use GI and maps often are unaware of how the GI or the map was influenced by such forces. One of the underlying reasons geographic information and maps are so powerful is that unless one knows a great deal about an area, it can be impossible to say specifically what cultural influences led to particular choices. Denis Wood demonstrates how cultural values connected to ideology become key parts of GI and maps that highlight a desired understanding of a portion of the world and subtly erase other parts of the world in the representation. Wood analyzed a highway map of North Carolina and showed that mapmakers do things like position explanatory text over poor areas of the state, for example.

Representing Other Cultures: Participatory GIS

A significant response to the selectivity of GI and maps arose in the late 20th century and has come to be known as participatory or public participatory GIS (PGIS). Multiple strands of development and also multiple emphases of activities can be grouped under the term PGIS. Some of the first prototypes of GIS-based analysis were developed by Ian McHarg in the 1960s with an explicit participatory concept in mind, but the concept of public participation geographic information systems arose mainly from broad academic discussions of GIS and society in the 1990s. PGIS arose out of attempts to move beyond social theoretical critiques to promote the development of GIS that could practically support the needs of communities.

The integration of GIS in a community can take on a variety of forms. One of the most common is the development of government-supported provision of equipment, experts, and training for local community groups. Sig-

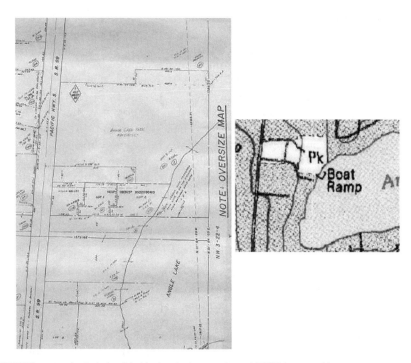

FIGURE 11.2. Angle Lake, Washington, in Assessor's and USGS topographic map representations.

nificant also is the development of sense-of-community GIS through local user groups, who work independently of any government group. These grassroots GIS may become very significant in the development of capabilities for local groups to develop and use existing GI and help foster a community sense of place, which has had important political, social, and economic consequences.

A key challenge for PGIS is the explicit integration of local knowledge. This enables powerful and insightful analyses, but runs the risk of making information available in forms that other people and groups may appropriate without accounting for the people who develop and nurture local knowledge, often over generations. For example, identifying the location of a site used for religious rituals may enable unscrupulous relic traders to find a prospective site and disturb or possibly destroy it, much as tomb raiders destroyed many Egyptian tombs.

PGIS, once implemented, can help local groups take on government functions. This can be a two-edged sword. On the one side, the use of GI and maps can be empowering and help communities cope with changes and make plans for the future. On the other side, the use of GI and maps often requires the training of experts who can limit access to these resources, in effect creating a new elite. The problems of incorporating GI and maps from people with various backgrounds can be especially challenging.

GI and Map Misuses

"All maps lie" is the statement made by the cartographer Mark Monmonier, drawing on the idea behind the classic book *How to Lie with Statistics*. Most cartographers would like to put this away as a simple "white lie" that masks the important and complicated work done by cartographers in making maps. As we have read earlier in the chapter, however, there are substantial differences between the distortions necessary to improve cartographic communication and the distortions that erase and obfuscate relevant things and events from the cartographic representation. We need to be able to distinguish between acceptable and unacceptable cartographic representations. In other words, we need to know when GI and maps are misused, or, in the extreme case, when GI or maps are propaganda.

When Is Distortion Propaganda?

Taking the definition that *propaganda* is information manipulated to fit particular ideological, political, or social goals with a malicious intent, then the distortion of any cartographic representation based on a person's or a group's ideological, political, or social goals makes for propaganda GI or maps. Examples are all too commonplace. Certainly, examples from Nazi Germany or the Cold War are blatant, but other examples can be drawn from advertising and political publications we find everyday. Following Denis Wood's profound unveiling of the power of maps, we may be inclined to consider all maps as propaganda, with exceptions possibly for purely technical maps of infrastructure. This type of relativism about the use and potential of propaganda GI and maps leads to negligence and disregard for the subtle, yet deceitful ways that propaganda GI and maps influence our understanding of things and events in the world.

Privacy Protection and Surveillance

In many areas of the world, increased surveillance has become much more commonplace and accepted. In many settings people have come to even find a degree of safety in the surveillance, but some people have begun to raise questions about the amount of data collected by government and private agencies and what happens with this data. As with information collected by product registration forms, applications, or using the phone or mail, which can be aggregated using common identifiers—for example, in the United States the Social Security number or house address—GI can also be combined to create more detailed analyses.

Most people would like to consider information about what and when they do things of everyday life including shopping, walking the dog, going for a run, and going to meet friends as private. New technologies make it easier to track movements and collect these kinds of individual-level information. Public safety concerns may make the collection of this information necessary for particular situations, but the widespread collection of individual data and its uncontrolled storage and use concerns many people. With these new technologies, many discussions around the world are unfolding about how to adequately protect privacy, yet at the same time ensure that adequate information is collected for law enforcement and defense purposes. The

---

In-Depth **What Is Propaganda?**

Propaganda, for most people, is the systematic manipulation of information to fit particular ideological, political, or social goals.

Thinking about the role and creation of maps in propaganda, we might want to start out with the insight that propaganda maps rely on distortions. Some are blatant and some are subtle, but these distortions, when critically examined, manipulate the GI or map to create a representation that sets out to convince that it, regardless of distortions, is the correct representation of that aspect of the world.

The quiet persuasiveness of GI and cartography is what makes it such a powerful form of deceitful manipulation. Packaged as effective communication, many people may not be able to assess the distortions. Ultimately, the propaganda of GI and maps have little to do with scientific aspects of geographic representation and cartographic representation, but are based in particular social and political interests advanced through the map.

FIGURE 11.3. A complex allegory from the early 20th century showing the Russian bear and the Russian army held back from by a stout barricade from the culture, farms, and industry of Finland.

From Kosonen (2000). Reprinted by permission of Katariina Kosonen.

legal and institutional decisions hinge very much on cultural conceptions of geographic representation and cartographic representation of individuals.

New surveillance technologies are pervasive and subtle. From drone aircraft that can stay aloft for extended periods to small computer chips the size of a grain of rice, the development of surveillance technologies is making it easier to record information about individual movements without the knowledge of the person under surveillance. Combined with GPS technology, it is becoming possible for many people to assemble data into a study of how individuals spend their time.

The first of these technologies in terms of relevance for recording individual movements and situations is satellite remote sensing. High-resolution remote sensing technologies offer resolutions of 61 cm, allowing for the detection of the presence of individuals in clear situations—for example, while standing on a concrete parking lot. Satellite remote sensing can be complemented and enhanced by ground-based remote sensing technologies. Britain has over 2.5 million surveillance cameras installed, offering many possibilities to record the movements of an individual. These technologies, which offer direct observations, can be easily combined with data from technologies that routinely collect information about the location and movements of people. The users of cellphones can be readily localized and phone records can be used to ascertain the general movements of a cellphone user, whether they are making calls or not. Other means arise from the increased use of electronic payment systems.

These technologies can be combined to create detailed profiles of individuals. This has been done for a number of surveys. The use for criminal investigation has been popularized in movies and TV shows. The mundane use of personal data to aggregate information for marketing and other purposes has received some coverage as the result of the increasing number of identity theft cases, but its use by telemarketers and direct mailers has lacked a clear response.

The complexity of an individual's movements and the limitations of the technologies leads to constraints. GPS receivers will not work inside buildings, and most high-accuracy remote sensing technologies will not record details about objects in the shadows of buildings or obscured by natural features (e.g., in ravines or under dense tree coverage). In other words, some unsurveyed areas remain, but as the technologies develop and people come to rely more on them for a sense of security, the cultural sense of what is appropriate to collect, record, and reuse about an individual's things and events will be changing.

Summary

Many choices of geographic representation and cartographic representation follow implicit cultural conventions. The success of GI or a map is to no small degree contingent on the ways that cultural conventions are included. The GI's or map's power is also a measure of how much of a contribution is made to the a culture's engagement with the world. Knowing the cultures of GI and maps can help make sense out of how GI and maps are used and manipulated, and how biases can become part of maps. In Western countries a central issue is accuracy—questions of what and how GI and maps represent. A second central issue is the question of why GI and maps are created. They are very expensive and must have good economic or political reasons. Finally, an engagement with cultural aspects should consider the choices people make. These points are relevant to ongoing debates surrounding surveillance.

Distortion is a common way to describe the inaccuracy of GI or maps. Propaganda is often accused of distorting the truth, but distortion is a necessary part of most GI and cartography. A helpful distinction comes from looking at the distortion of propaganda materials in terms of a malicious intent or willful distortion. This provides no black-and-white criteria to distinguish propaganda, but still can help identify propaganda in practice.

In-Depth **Different Ways to Represent Geographic Knowledge**

Textbooks like this book focus on the creation and use of GI and maps for the representation of geographic knowledge, but we certainly should at the bare minimum mention other forms of representation: text, pictures, drawings, and verbal descriptions.

Text is perhaps the most common form of geographic knowledge representation because it is easier to create than GI and maps, but is obviously limited. However, as we all know from experiences giving and getting directions, text is often far more convenient and easier to use than a map drawn by a friend or colleague. For particular purposes, it may even be easier to use than a generic highway or city map.

Pictures may look like maps, but often simply use one or two conventions to help readers make sense of the map. Maps used in advertising or at tourist destinations often are really just pictures that may add a legend and standardized symbols.

Drawings are a third type and are very common, but usually limited to communicating geographic knowledge related to a particular purpose—for example, comparing the size of two shops, streets, or countries. Drawings may loosely rely on conventions of mapping, but are rarely consistent due to the limited scope of their use.

Indigenous cultures often rely on verbal descriptions, not just for communicating where things are and how to get to places, but to share complex and important stories about the culture's creation and meanings associated with places. These meanings can reflect generations of experiences, replacing scientific observations and measurements shown on a map with a deep lore and understanding of place.

In-Depth Placenames and Conflicts

The names given to a place may change over time and may reflect deeper changes in local society and culture. However, usually only one name is presented on a map. Most maps indicate the capital of Italy and location of the Vatican City with the Italian name, Roma, but some may replace it with their language's name—for example, in English, Rome; in Polish, Rzym; and in German, Rom. This may be common practice in a country's schools and in the general media, but a globalizing world has meant that people need to be more attentive to naming practices.

This has long been the case in multicultural states. For example, in Switzerland, where four languages are nationally recognized, cities in zones with influences from different cultures may often be identified on maps with two names—for example, Delsburg and Delemont or Neuenburg and Neuchâtel. Cities clearly in one language area will usually be shown with only one name, although exceptions occur. These exceptions can be laden with conflict and be irresolvable. For example, the large lake that Geneva lies on is known there as Lake Geneva, but along the majority of the shoreline in Switzerland and in neighboring France, the lake is known by the name of Lac Leman, which derives from the Roman name the lake, Lac Lemanus. To this day, Geneveans and the international community use their name, while others use Lac Leman.

Many such conflicts occur around the world. Perhaps one of the most significant geographic naming conflicts is between Japan and Korea over the name of the sea between the two countries. The sea was known variously as the "Korean Sea" or the "Japanese Sea" for most of the modern period. When

Japan began expanding into Manchuria and Korea in the 20th century, the name became the now commonplace "Sea of Japan." For sometime now Koreans have been seeking to have the name changed officially by the UN to reflect the historical uses, but although their claim has been acknowledged, its recognition and adoption by the UN has been slow in coming.

In-Depth Crises of Representation: Crises for Geographic Information and Maps?

A number of social philosophers have criticized the representational culture of Western modernity since the late 19th century. Their impact has been substantial and broad. For GI and cartography, their endeavors to understand the activities of representation and construction of GI and maps have led to a number of contributions, which have begun more and more to influence scientific and professional cultures. In his recent book, *A History of Spaces*, John Pickles engages these issues by developing three crises of representation that I interpret here in the context of broader crises of representation.

The first of these crises relates to the assumption of objectivism in GI and maps. Many GI and map makers and users presume that their GI or maps approximate the real world as a correspondence or true relationship between symbols on a map or points, lines, and areas with a single "reality." For them, the making of GI and maps involves the straightforward collection, preparation, transmission, and reception of information. In other words, a good map is one in which the information intended for communication by the maker corresponds to the information received by the map user. The map, in objectivism, is then an accurate representation of the real world. Many people, of course, have come to realize that the accuracy of GI or maps has as much to do with what people are trained to see and measure. Out of the crisis of objectivism, people recognize that GI and maps are as bound by conventions and ideologies as they are by the science they deploy or follow.

The second of the crises relates to the assumption that what GI and maps represent is natural. Clearly, a corollary of the first crisis, many people who wish to believe that maps correspond to what can be found in nature have ended up finding out that any such correspondence results from the role of maps in first creating what is to be found—for example, wetlands or low-density housing. The boundary between the scientific and the political roles of maps is very fuzzy and for some any attempt to draw a boundary is inadequate. The intentions of GI and map makers and users matter. For David Turnbull, this crisis culminates in the recognition that maps precede territories, creating the spaces they map and then claiming them to be natural.

The third crisis is that of subjectivism. The knowledge represented in a map is always connected to forms of social interest. Biases in apparently "objective" GI and maps result. This crisis involves the recognition that we always must consider the interests of GI and map users and makers when considering the cultures and roles of GI and maps in any society.

Review Questions

1. When is a map considered to be propaganda?
2. Why do maps and GI figure so significantly in our understanding of the world?
3. What are the three types of culture affecting GI and maps?
4. How can cultural issues and values influence maps and GI?
5. How have civilizations used maps?
6. What is an example of an indigenous form of mapping?
7. What is the main motive behind Participatory GIS?
8. Why are people increasingly concerned about privacy protection and surveillance?
9. How can the privacy of an individual be impinged by surveillance technologies?
10. What is the difference between distortion and propaganda uses of GI and maps?

Answers

1. When is a map considered to be propaganda?

   A map should be considered to be propaganda when the cartographic representation, geographic representation, or communication is malicious in intent.

2. Why do maps and GI figure so significantly in our understanding of the world?

   Maps are powerful visual ways to acquire information about places, things, and events we might never directly experience.

3. What are the three types of culture affecting GI and maps?

   The three types are national culture, indigenous culture, and disciplinary culture.

4. How can cultural issues and values influence maps and GI?

   They affect how people make meaning from GI and maps.

5. How have civilizations used maps?

   To the best knowledge of archeologists, all civilizations have used maps to represent geographic things, events, and the relationships between them.

6. What is an example of an indigenous form of mapping?

   Stick charts used by Pacific South Sea islanders are one example.

7. What is the main motive behind Participatory GIS?

   Participatory GIS seeks to support the needs of communities

8. Why are people increasingly concerned about privacy protection and surveillance?

   Private companies and governments are collecting and combining more and more information about individuals.

9. How can the privacy of an individual be impinged by surveillance technologies?

High resolution remote sensing technologies could be used, for example, to detect the location of people at outdoor facilities.

10. What is the difference between distortion and propaganda uses of GI and maps?

Distortion reflects the unavoidable inaccuracy of geographic and cartographic representation to the actual situation. Propaganda distorts with malicious intent.

Chapter Readings

Cosgrove, D. (2001). *Apollo's Eye: A Cartographic Genealogy of the Earth in the Western Imagination.* Baltimore: Johns Hopkins University Press.

Craig, W. J., T. M. Harris, et al. (Eds.). (2002). *Community Participation and Geographic Information Systems.* London: Taylor & Francis.

Ghose, R., & S. Elwood. (2003). Public Participation GIS and Local Political Context: Propositions and Research Directions. *URISA Journal, 15*(APA II), 17–24.

Harley, J. B. (1989). Deconstructing the map. *Cartographica, 26*(2), 1–29.

Kosonen, K. (2000). *Kartta ja kansakunta: Suomalainen lehdistökartografia sortovuosien protesteista Suur-Suomen kuviin, 1899–1942* [The map and the nation: Finnish press cartography from the protests of oppression years to the images of Greater Finland 1899–1942]. Helsinki: Suomalaisen Kirjallisuuden Seura.

Monmonier, M. (1991). *How to Lie with Maps.* Chicago: University of Chicago Press.

Monmonier, M. (1993). *Mapping It Out: Expository Cartography for the Humanities and Social Sciences.* Chicago: University of Chicago Press.

Monmonier, M. (1995). *Drawing the Line: Tales of Maps and Cartocontroversy.* New York: Holt.

Pickles, J. (Ed.). (1995). *Ground Truth: The Social Implications of Geographic Information Systems* (Mappings: Society/Theory/Space). New York: Guilford Press.

Pickles, J. (2004). *A History of Spaces: Cartographic Reason, Mapping, and the Geo-Coded World.* New York: Routledge.

Rundstrom, R. (1995). GIS, Indigenous Peoples, and Epistemological Diversity. *Cartography and Geographic Information Systems, 22*(1), 45–57.

Sieber, R. E. (2000). Conforming (to) the Opposition: The Social Construction of Geographical Information Systems in Social Movements. *International Journal of Geographic Information Science, 14*(8), 775–793.

Thrower, N. J. W. (1972). *Maps and Man: An Examination of Cartography in Relation to Culture and Civilisation.* Englewood Cliffs, NJ: Prentice Hall.

Toledo Maya Cultural Council and Toledo Alcaldes Association. (1997). *Maya Atlas: The Struggle to Preserve Maya Land in Southern Belize.* Berkeley, CA: North Atlantic Books.

Turnbull, D. (1989). *Maps Are Territories: Science Is an Atlas.* Chicago: University of Chicago Press.

Turnbull, D. (1998). Mapping Encounters and (En)countering Maps: A Critical Examination of Cartographic Resistance. In *Knowledge and Society* (Vol. 11, pp. 15–43). London: JAI Press.

Wood, D. (1992). *The Power of Maps.* New York: Guilford Press.

Web Resources

The Integrated Approaches to Participatory Development (IAPAD) website offers a number of good resources and pointers to additional resources and discussions. See *www.iapad.org/*

The Aboriginal Mapping Network offers a number of good mapping resources. See *www.nativemaps.org/*

The British Library has an online exhibit that engages the role of propaganda maps. It is available online at *www.bl.uk/whatson/exhibitions/lieland/m0-0.html*

Chapter 12

Administration of Spaces

This chapter covers the key roles of administrative GI and maps in creating and maintaining the world as we know it, both globally and locally. Defense, commerce, and taxation are key government activities that require or benefit from clearly conceived and communicated GI and maps. Maps and geographic information are involved in almost every government activity.

After considering the impact of culture on GI and maps in Chapter 11, we can see that geographic information and maps for administrative activities and governmental policies are important. In a representative democracy, many social relations are regulated politically; thus, those social relations can be represented in the form of geographic information and maps, which means that geographic and cartographic representations also have a political character. Even attempts to maintain political "neutrality" in mapmaking can still have a political character in situations where there is no legitimate "politically neutral" point of view. An attempt to remain neutral here possibly nullifies one of the principal roles of geographic representation, which is to help policymakers in making better decisions.

A second political dimension of geographic representation is that we "create" nature in the choices we make for representing natural objects. Deciding how to represent nature—and how to define it—is an important political activity that often involves the administration of spaces. From spotted owl protection to administrative regulations regarding the definition of protected plants and species, the geographic representation and cartographic representation of the "natural" world plays key roles in how people understand and act in the world.

Chapter 11 examined the different cultures and misuses of geographic representation and map representation; this chapter turns to cover the empowering role and uses of GI and maps for administrations. It starts with an overview of how GI and maps have become key references for the organization of all social and economic activities. Administrative, especially governmental, maps have created the world we know and continue to be key parts of how we understand and engage that world. The chapter next considers the significance of land ownership recording—the cadastre—for capitalist economies and societies. The third section examines how administration activities have influenced the development of GI technologies. The closing section examines various sources of geographic information.

Administration of the World We Know

Through geographic and cartographic representations governments create spaces. The external boundaries and internal administrative divisions of the United States, Great Britain, Australia, Canada, Germany, Poland, India, and every other country have been created by many government administrations. Laws, regulations, and procedures rely on coordinate and locational systems.

Administrative activities in the 50 U.S. states provide both positive and negative examples of map use. From the settlement on the Eastern Seaboard and along the southern and western coasts by Europeans to relentless attempts to create advantageous election districts through gerrymandering, examples from the United States highlight the role geographic and cartographic representation play in our lives.

The history of the settlement of the colonies on the Eastern Seaboard shows that Europeans worked to create the illusion that the land they were settling was relatively "unoccupied." As William Cronin describes the development of European settlements in *Changes in the Land*, setting boundaries for private land ownership and creating villages that could be readily mapped (in contrast to indigenous practices of sharing knowledge mainly through narratives and rituals) was a key part in establishing European administrations in North America.

The westward expansion of the United States points to the political importance of administrative activities that relied on geographic and cartographic representations. By failing to require surveyors of the lands brought under U.S. control to record existing indigenous habitation, early geographic representations rarely showed existing settlements and land use. This geographic representation suggested that vast areas of land were vacant and freely available.

The sharp angles of roads and agricultural land in these parts of the United States have always been interpreted from various points of view. For some people, they indicate the rational underpinnings of the American economy and society. For others, the geometrical abstractions further remove traces of indigenous communities and much of the natural attraction of the land, subordinating it to calculative economics of gridded space.

Finally, the complexity of U.S. governmental administration results in short-term consequences being given priority over long-term. In the hierarchy of public administration across the 50 states, there are some 3,200 counties, further subdivided into 31,000 special departments (e.g., water, sewage, trash collection, fire protection). This makes local government in the United

States very hard to navigate. Resolving overlapping jurisdictional responsibilities has become an important obstacle in addressing many issues in the United States, most recently homeland security.

The Cadastre: Recording Land Ownership and More

Capitalist economies rely on the cadastre, which makes it one of the more important administrative divisions of spaces. The cadastre, a term that originated in Latin, refers to the registry of land ownership. In most places today the cadastre is fundamental to determining the rights and responsibilities of landowners and land users.

The cadastre consists of two parts. One part records the ownership of land in the form of a legal title, which also defines the land property. The second part describes the boundaries of the land property. The two parts are usually separately maintained—opening the door to substantial differences. This means that the land shown in a cadastral map need not match the land described by a cadastral title.

As a formal record of landownership, the cadastre is an important tool in both governmental administration and private enterprise. It serves to help

FIGURE 12.1. The boundaries of 293 public administration units in the Minnesota Twin Cities area.

Courtesy of MetroGIS, St. Paul, Minnesota.

governments identify who has to pay taxes, and what amounts, on their land-ownership, and also indicates the rights, responsibilities, and obligations of landowners and users. It is used by banks that provide mortgages, lawyers who draft contracts for land transactions, and companies that want to determine how land prices are changing in an area. In these ways, the cadastre is just one part of the broader socioeconomic uses of land called "land tenure." The Food and Agricultural Organization of the United Nations (UN) defines *land tenure* as "the relationship, whether legally or customarily defined, among people, as individuals or groups, with respect to land" (See the FAO discussion "What Is Land Tenure?" in this chapter's Web Resources.) In contrast to the explicitly formal role of the cadastre, land tenure also includes individual agreements, such as when two neighbors agree to use parts of both properties as a shared driveway and place to park their cars, or when an older person in the family allows younger family members to use the property without charging them money or selling the property to them. It also involves the more complex relationships between people that involve land.

The cadastre records landownership and through its formal significance accomplishes or certainly helps to accomplish much more, but it is only part of the relationships people have to the land they live on or use. The cadastre is a key part of any capitalist economy, but its relevance depends to a great extent on its tangibility for day-to-day relations.

In the United States, most landownership is recorded using the Public Land System (PLS) (see Chapter 5 for more specific details). This cadastral system guarantees a relatively simple system of land title and boundaries, but errors and complex situations do often arise. Many states in the United States require title insurance on all land transactions to cover the risk that

FIGURE 12.2. Portion of an assessor's map.

FIGURE 12.3. Modern cadastre in Germany.
Courtesy of Dirk Linnemüller.

some other valid claim might be made related to an area of land. In this system any maps or plans showing the boundaries to the title are incidental.

On the East Coast and in a few other areas of the United States (Texas, Louisiana, and parts of California and New Mexico) the metes-and-bounds system dominates. This system is more involved for surveyors and can be very expensive if differences between neighboring property owners must be resolved in court. The Torrens system, developed by Sir Robert Torrens in Australia in 1857, provides for a clear registration of property and creation of a title. The title does not transfer to later owners; it must be reregistered with the court. In this system maps and plans are integral to the title.

Administration Impacts on Geographic Information Technologies

Government administrations have had and continue to have significant impact on the development of geographic information and maps, particularly through their role in developing new technologies. Obviously, the complexity of cadastres means attempts should be made to improve administrative technology. Many of these efforts have focused on internal improvements and are scarcely noticed by the general public, except when permits and taxes become more transparent. These internal improvements have focused on two areas: improving administrative activities through the use of GI technologies and improving administrative coordination between different agencies. We will examine some of the successes in improving administrative activities in Chapter 13. In this chapter, we focus on how government coordination has been improving because of administrative-led developments of GI technologies.

Multipurpose Land Information System

Because improving administrative activities can be greatly aided by improving data sharing, many developments have focused on improving data sharing, not only through technological standards, but by establishing a common reference base for administrations. The Multipurpose Land Information System (MPLIS) is perhaps the best known and most significant of concepts for improving administrative data sharing.

MPLIS has its roots in cartographers' attempts to develop multipurpose cartography in the 1950s. *Multipurpose cartography* relied on the preparation of different thematic layers using a common coordinate system that facilitated the manual overlay of the layers in the preparation of the printed map. The thematic layers corresponded to administrative activities—for example, U.S. states as a base layer combined with another layer of highways or a layer showing railways to produce maps showing different parts of the transportation infrastructure.

MPLIS starts with this multiple use concept and simultaneously extends it and focuses it on local governments, counties, or municipalities. The extension of the multipurpose cartography concept focuses on making the resulting combined information the catalyst for administrative coordination. This is possible because the GI technology developed since the 1960s made it very easy to combine data as long as a common coordinate system was used. To make presentation maps a cartographer was still called for, but maps for administrative analysis or straightforward informational purposes could now be produced by people working in administrative offices instead of specialized cartographers.

For the MPLIS to work, a great deal of coordination and data sharing is required. Administrative resistance is most often the greatest stumbling block for data sharing. Conflicts between different mandates and different information needs often impair the development of the MPLIS in local governments. The many success stories around the world point to the great significance of this administrative concept for the improved efficacy of administrative cartography and GI.

Spatial Data Infrastructures

In some senses spatial data infrastructures (SDI) are extensions of the core MPLIS concepts (coordination and data sharing using GI stored in a common coordinate system) to other levels of government. The first, and still most significant, development of the SDI occurred in the United States through the creation of the National Spatial Data Infrastructure (NSDI).

Extending MPLIS concepts to larger geographic regions, the basic principle of the SDI is that freer access to administrative GI contributes to good governance and a better society, government, industry, and people. The MPLIS provided evidence for the potential; the SDI grapple with the issues of developing the capabilities of networked data sharing and administrative coordination that use computer networks to transfer data between different sites.

Because the NSDI was conceived while the Internet was being developed, its underlying concepts work without a network; its implementations explicitly rely on the Internet. The initial goals for the NSDI focused on civilian federal agencies and emphasized the importance of these agencies sharing GI as a means to reduce the expenditures for the federal government. Although the NSDI was conceived of as guidelines for the development of infrastructures that reached from federal agencies to local administrations, it was only sporadically implemented as intended because the NSDI lacked legislation that broadly supported the involvement of local governments or required the participation of state and local administrations. On this note, it is important to recognize the inspiration and conceptual guidance the NSDI has provided administrations in the United States and other countries. While perhaps not realized as intended, the NSDI has been an important driving force for improving interoperability.

The NSDI in the United States consists of three components: framework, vertical and horizontal dimensions, and the availability of free or low-cost data. The *framework* defines seven data layers needed for nearly every government activity. This data should provide nationwide coverage. The framework has both vertical and horizontal dimensions. *Vertical* refers to data exchange and sharing between different levels of government; *horizontal* refers to the exchange and sharing between administration units at the same level of government. Finally, the NSDI includes the concept that framework data should be *freely available*. Other data can be freely available, or may be only available after paying a fee or registering with the data provider. This theoretically allows for the greatest reduction of data collection and maintenance costs by providing support for as many different GI applications as possible.

Other SDI were developed after the U.S. NSDI was proposed. They directly incorporated networking capabilities in their design and implementation. These SDI also are largely developed to support primarily intragovernmental data exchange and sharing. Public access is generally quite limited. Most common is to allow access to Internet services that facilitate the creation of maps. Some SDI are less integrated in the NSDI. They offer services to examine descriptive data about available data sets (metadata) or, after registration, to log on to a protected website and download data.

Free, simple data sharing is the central concept for the SDI, but has been greatly impaired in every instance by financial and participation issues.

Data sharing is made difficult by problems allocating and distributing costs, but institutional and political issues related to participation play a sizeable role as well. While there is no conclusive evidence that charging for GI provides enough revenue to recover the costs and maintenance of the GI, nor even the costs of managing the system for regulating use and charging, most administrations around the world continue to charge, even in cases of use by other administrative agencies. People voice concerns about paying for the data necessary to assure public safety as well as civil and environmental protection and administrative budgetary concerns, but the developments of SDI have been greatly impaired by cost recovery. SDI remains one of the key concepts guiding developments of GI. How individual regions and countries address the challenges is an important question for the future of GI and cartography.

Digital Libraries

Work on digital libraries addressed a very important source of information for society: public libraries remain great repositories of information and offer support for knowledge economy activities that find little support elsewhere. Compared to traditional map libraries, digital map libraries offer some capabilities that are of great benefit to many users of GI and maps. The key difference is that digital map libraries combine both paper maps and GI.

This can occur through the scanning of existing paper maps (when possible) and their storage in a digital format. Even with referencing to a coordinate system, scanned paper maps are still cumbersome in comparison to GI. However, when copyright laws allow it, scanning provides a straightforward way to collect information. Digital libraries of GI also support novel methods for accessing GI that reflect the complexity and the variability of how people work with and use GI and maps. Digital libraries can distribute the physical storage of GI to different sites. The possibility also exists of accessing information from multiple sites for digital libraries. A person accessing a university digital library of GI and maps may transparently access GI and maps from other university libraries.

Digital Earth

Proposed in the late 1990s and recently developed as a commercial software application by Google, Digital Earth permits a person anywhere in the world to access GI for any place on the earth, at variable scales and resolutions, via the Internet. Its relevance for administrations working with GI are limited, but it certainly is important for agencies involved in global issues and for providing three-dimensional visualizations of many areas of the earth.

Digital Earth also exists as a specification for the global georeferencing of GI. This mainly supports global environmental research, but it is clearly a useful reference framework for a much larger number of uses.

Research Support

All these administration activities have some level of research support that goes along with the actual administrative developments. In general, a relatively small amount of research support, in the case of the MPLIS and NSDI, for example, has been the seed of wide-ranging administrative developments. Governments also specifically fund research on improving data collection and maintenance techniques, assessing and improving administrative services, and developing new technologies and approaches to make better use of GI and maps. This support has been crucial at various points in the development of the over U.S. $200 billion/year GI and cartography industry worldwide.

Government Sources of Geographic Information

Most available GI comes from government agencies. Sometimes it is not easy to find GI for a specific region. This section only attempts to overview key sources. For specific GI needs, you should first contact a nearby map library.

North America

The United States seems to stand out because of the widespread ease of obtaining data, but that is actually certain only for most data held by federal civilian agencies. Public domain data collected by civilian agencies of the U.S. government with general funds are considered to belong to U.S. taxpayers. All states have their own regulations, generally called "open records laws," which describe which data can be made available, under what restrictions, and at what costs.

The U.S. National Atlas (*www.nationalatlas.gov*) is a good starting point. It supports the online creation of maps as well as data downloading. Data about roads, address ranges, and census geography is available at *http://www.census.gov/geo/www/tiger/*; however, this data requires some processing before it can be viewed and no online browser is available at this site. Some remote sensing imagery can be found at *http://www.class.noaa.gov/nsaa/products*. Additional remote sensing data and individual images are available at *http://eros.usgs.gov/archive/nslrsda/*.

The U.S. states maintain individual websites for getting information and accessing GI and making online maps. There are far too many to list here. You may wish to go to the Geospatial OneStop for a basic list and access information (*http://www.geo-one-stop.gov/*). Local governments also may maintain online access possibilities. You should be able to find these out by going to the home page of the local government (county or municipality) in question.

In Canada, the Canadian National Atlas (*http://atlas.nrcan.gc.ca/site/english/index.html*) is an excellent resource, as are increasingly the provincial, county, and municipal governments. GI for Mexico is far scarcer.

Europe

The GI available in Europe tends to be only available for government agencies or only after purchasing. A good starting point for European environmental data is the UNEP-GRID office in Geneva (*http://www.grid.unep.ch/data/index.php*). Another good source is the European Environment Agency's data service (*http://dataservice.eea.eu.int/dataservice/*). At this site you can browse metadata and if you sign an agreement you can also download data.

The sources for GI from various countries and regions vary greatly. For instance, Danish GI is available online at *http://www.grid.unep.ch/data/index.php*, but it must be purchased. The U.K. Ordnance Survey has a great deal of GI, which can be accessed after applying and usually paying a fee (*http://www.ordnancesurvey.co.uk/oswebsite/*). Poland makes some data available over the Internet for viewing only (*http://217.153.152.212/bdo/*). Other sources can be found at the European Umbrella Organization for Geographic Information (EUROGI) website (*http://dataservice.eea.eu.int/dataservice/*).

Other Parts of the World

Some GI for New Zealand is available (*http://www.geographx.co.nz/*). Other areas of the world have a great deal of GI, but most is part of global or regional data sets. You can try the GI clearinghouses (*http://www.fgdc.gov/clearinghouse/clearinghouse.html*) or the Geography Network (*http://www.geographynetwork.com/*).

Summary

Mapping and GI empower the creator and user. Mapping is part of almost every government activity because of its power. Through geographic representations and cartographic representations governments create spaces. The cadastre is a prime example of how administrations in Western civilization have created spaces. No matter how much effort the administration puts into a cadastre, it must be accepted that its boundaries are related to actual boundaries to be relevant. The Multipurpose Land Information System (MPLIS) was one of key post-World War II information technology concepts for administrating local government activities. The Spatial Data Infrastructure (SDI) develops these concepts to involve multiple governmental units and benefit from computer networking. Other concepts that have advanced the ways of administrating spaces are digital libraries and the Digital Earth. The results of these concepts can be seen in the increasing access to GI.

Review Questions

1. Why is there such a large administrative interest in GI and maps?
2. What are the two ways of recording the boundaries of land owner-ship?
3. What does the abbreviation PLS stand for?
4. What is the basic principle behind the SDI?
5. How do coordinate and locational systems become key references?
6. What is the definition of *cadastre*?
7. What makes data sharing so difficult?
8. What does "public domain GI" mean?
9. Explain the key principles of the Multipurpose Land Information Sys-tem?
10. To which parts of government in the United States does the public domain principle apply?

Answers

1. Why is there such a large administrative interest in GI and maps?

   Defense, commerce, and taxation are key government activities that require or benefit from clearly conceived and communicated GI and maps.

2. What are the two ways of recording the boundaries of land owner-ship?

   The most common approaches to recording the boundaries of land owner-ship are metes-and-bounds surveys and systematic surveys (e.g., PLS).

3. What does the abbreviation PLS stand for?

   PLS stands for Public Land Survey. "System" is often appended to create the abbreviation PLSS.

4. What is the basic principle behind the SDI?

   The basic principle of the SDI is that freer access to administrative GI bene-fits good governance and society, government, industry, and people.

5. How do coordinate and locational systems become key references?

   By becoming part of administrative laws, regulations, and procedures, coor-dinate and locational systems become part of social interactions and help structure the human-influenced geography of an area.

6. What is the definition of *cadastre*?

   Cadastre is a Latin term that refers to the registry of land ownership. In most places today that understanding is extended to include the role of ensuring the rights and responsibilities of landowners and land users.

7. What makes data sharing so difficult?

   Data sharing is directly made difficult by problems allocated and distribut-ing costs, but institutional and political issues play a sizeable role as well.

8. What does "public domain GI" mean?

Data collected by civilian agencies of the U.S. federal government with general funds that are considered to belong to U.S. taxpayers are generally available for no or little charge.

9. Explain the key principles of the Multipurpose Land Information System?

The Multipurpose Land Information System (MPLIS) places the GI from county or municipal agencies using a common coordinate system. Some systems additionally use the parcel, or cadastral, "layer" as the base layer to align other layers.

10. To which parts of government in the United States does the public domain principle apply?

The public domain principle in the United States only applies to civilian federal governmental agencies. State and local governments follow the open records laws of their state.

Chapter Readings

Cho, G. (2005). *Geographic Information Science: Mastering the Legal Issues*. New York: Wiley.

Chrisman, N. R., & B. J. Niemann. (1985). *Alternative Routes to a Multipurpose Cadastre: Merging Institutional and Technical Reasoning* (AutoCarto 7). Washington, DC:

Cronon, W. (1983). *Changes in the Land: Indians, Colonists, and the Ecology of New England*. New York: Hill & Wang.

National Research Council. (1980). *Need for a Multipurpose Cadastre*. Washington, DC: National Academy Press.

National Research Council. (1990). *Spatial Data Needs: The Future of the National Mapping Program*. Washington, DC: National Academy Press.

National Research Council. (1994). *Promoting the National Spatial Data Infrastructure through Partnerships*. Washington, DC: National Academy Press.

Tulloch, D. L., B. J. Niemann Jr., et al. (1996). *A Model of Multipurpose Land Information Systems Development in Communities: Forces, Factors, Stages, Indicators, and Benefits* (GIS/LIS '96). Denver, CO: ASPRS/AAG/URISA/AM-FM.

Sherman, J. C., & W. R. Tobler. (1957). Multiple Use Concept in Cartography. *Professional Geographer, 9*(5), 5–7.

Web Resources

For a German document presenting the role of GI and maps in governance and administration, see *www.imagi.de/en/download/Geoinformation_broschuere_engl. pdf*

For an FAO discussion of land tenure, see *ftp://ftp.fao.org/docrep/fao/005/y4307E/y4307E00.pdf*

For an adaptation of the MPLIS concepts for the U.S. Bureau of Land Management, see *www.blm.gov/nils/*

A starting point for information about the U.S. NSDI is available online at *www.fgdc.gov/nsdi/nsdi.html*

Information about individual U.S. state open record Laws is available online at *http:// foi.missouri.edu/citelist.html*

The European Commission's INSPIRE sets out to establish an SDI for environmental programs. See *http://projects.jrc.cec.eu.int/show.gx?Object.object_id=PROJECTS000000000001AD63*

The Global Spatial Data Infrastructure (GSDI) is an organization promoting the development of SDIs around the world. See *www.gsdi.org/*

Information about Digital Earth concepts is available online at *www.digitalearth.gov*

The conceptual heir to the Digital Earth concept is Google Earth, available online at *http://earth.google.com*

For an example of an environmental study's use of GIS in Columbia, see *www.lwr. kth.se/Publikationer/PDF_Files/LWR_EX_05_19.PDF*

The Canadian National Atlas offer a cornucopia of Canadian data. It is available online at *http://atlas.gc.ca/site/english/index.html*

Exercise

1. *Why Should GI Be Expensive?*

GI can be hard to obtain and expensive. Why is that? Consider both user and provider perspectives.

Should GI be made as cheap as possible or should it be sold for enough money to cover the costs of collection and maintenance?

2. EXTENDED EXERCISE **Geographic Information on the Web**

Overview

Much of the information represented in maps can now be found on the World Wide Web. In addition to presenting reproductions of maps, many websites also offer abilities to produce "customized" maps.

Concepts

Data on the web, some say, is replacing traditional maps. As you work through this exercise think about that statement and about the advantages and disadvantages of maps available on the web.

Part 1: Using Online Data to Make Maps

STEP 1

In this part of the exercise you will work with information from several websites. At the U.S. National Atlas site make a map of Minnesota showing either demographic or environmental characteristics.

U.S. National Atlas: *http://nationalatlas.gov/natlas/natlasstart.asp*

Follow the instructions on this webpage and produce a map of Minnesota or the Upper Midwest for an agricultural attribute, such as Soybeans for Beans—1997.

QUESTIONS

1. What are the minimum and maximum or first and last legend values?
2. What kind of measurement framework (nominal, ordinal, interval, ratio) is used for the data?

STEP 2

Make a map of Minnesota at this site: *http://mapserver.lmic.state.mn.us/landuse/*

Choose a county from this webpage. After choosing a county, generate a table of statistics by choosing the Create Statistics *radio button and a county.*

QUESTIONS

3. What is the dominant land use in the county and what is its name?
4. What percentage of the county is made up by urban and rural development?

5. Is there any mining in the county?

Scroll to the bottom of the page and use the Create multicounty land use and cover statistics link. You will then be at the Minnesota land use and cover statistics page and can prepare statistical summaries for answering the next questions.

6. What Minnesota county has the greatest percentage of forested land?
7. What Minnesota country has the smallest acreage of water?

STEP 3

Continue now to look at some demographic data for Minnesota. Go to *http://www.lmic.state.mn.us/datanetweb/php/census2000/c2000.html*. This brings you to a web-page listing a variety of statistical data available for Minnesota. Choose Demographics from the menu box on the left and then the Census Reports and Mappings link. Under the heading Population Profiles choose Population in 1970, 1980, 1990, 2000 and then click the Mapping button on the right. Enter a title for the map (e.g., "Minnesota Population in 1970"). Click the Define legend button to make your own legend following this example.

STEP 4

Now, let's take a look at an example of a remote sensing application. Go to *http://earthshots.usgs.gov/Wyperfeld/Wyperfeld* and describe what this imagery shows.

QUESTIONS

8. What has been happening in Wyperfeld?
9. How many of the 27 seasons were captured by the Landsat satellites? (Hint: read the whole story)

Part 2: Find Data

In this part of the exercise, you can search for data for any area in the United States you are interested in. Suggested starting points are:

www.mapsonus.com
www.mapquest.com
www.weather.com

Using the site of your choice, print the map you make and answer the following questions on a separate sheet of paper.

QUESTIONS

9. What is the URL you went to?
10. What kinds of maps are available at this website?
11. Can you interactively create (choose characteristics or attributes, your own legend values, etc.) your own maps at this site?
12. Describe the map you made: What does it show? Does it look like what you would have expected? What is the legend?

Part 3: Take a Bad Map . . . And Make It Better

This is the most creative part of the exercise. You will need to find and print a map that you find "bad" for any number of reasons: the colors, the legend, the theme, the biases (projection, scale, symbols), or whatever. You might want to first go back to the lecture on misuse of maps and map propaganda to see a few examples of "bad" maps. These questions also relate to Campbell's discussion of cartography in Chapter 1.

QUESTIONS

13. Why is this map horrible?
14. What could be done to improve it?
15. Geographic information on the web, some say, is replacing traditional maps.

After completing this exercise think about whether there advantages to traditional paper maps that you think web-based maps will not be able to match?

Part IV

GI Analysis: Understanding Our World

Chapter 13

GI Analysis and GIS

Every day people make spatial choices based on geographic information analysis: how to drive home, what is the best spot for fishing, where to go on vacation, where is the most convenient childcare, which site is preferred for opening a store. Looking only at a map will not necessarily provide you with the best choice. It can help, but reading a map is time-consuming and the map may be inaccurate. You often need fuller comparisons and analysis of data to make the best choice. Indeed, geographic information analysis is becoming more common through online and software navigation applications. Geographic information analysis also has been and remains a key part of many government and business activities. These navigation applications and professional analyses are similar in concept, but the commercial orientation of many online and commercial software applications hides many of the details of their analysis and restricts consumers. Professional analysis almost always involves considering many details and choosing between possibilities.

In terms of geographic information, analysis takes many forms. It can involve little more than the comparison of two data sets collected at different times for the same area. Or it can be complex: the analysis of the relationship between existing residences and a proposed highway that will use sound buffers—entailing complex geostatistical analyses that dynamically model the processes of noise being created by different types of traffic through the course of a 24-hour period. Geographic information analysis involves the explicit or implicit translation or transformation of things and events into patterns and processes.

This chapter begins with an overview of this translation or transformation. Translations and transformation take place in the context of communication. Considering the communicative role of the analysis helps one to get a grasp on the different forms of geographic information analysis. The chapter next provides an overview of geographic information analysis types and GIS. This chapter concludes with a discussion of GIS definitions and some examples of how GIS is used for analysis in environmental and urban domains.

Analysis for Communication

Ultimately, all analysis is used for communication. The history, culture, and purpose of any analysis will vary from internal sketches and plans all the way to advertising banners and TV presentations. Underlying the choices for analysis are many of the issues discussed in Chapters 3, 9, and 10. In a nutshell, successful communication relies on appropriate geographic representation and cartographic representation. For example, If an analysis needs to be made of a health clinic's accessibility in a large metropolitan area, it will likely be necessary to consider most roads in the area, but not represent them in a map of the entire area comparing accessibility between clinics. How choices are made is an important part of every geographic information analysis.

The choices reflect the conventions of the people preparing the analysis and the cultural values of the people who will be using the results. People familiar with the clinics' locations may leave off information about smaller streets in the cartographic representation that is obvious to them, potentially confusing people who are new to the area or who have never been to the clinic before. But even before the cartographic representation is prepared, the geographic representation will already reflect certain conventions. The measurements, observations, and relationships of the clinics are strongly influenced by the perceptions, backgrounds, and disciplinary perspectives of the people involved in preparing the analysis.

Issues for Geographic Information Analysis

In particular, for geographic information analysis, three key issues come to mind. These issues often form trade-offs, not necessarily absolutes or either's and or's, but each issue involves finding different balances between analysis and communication.

PATTERNS/PROCESSES

One of the most fundamental choices is deciding how to analyze the relationships involving things and events. Most GIS software only supports the storage of things as patterns. Events can be modeled as processes, with the model aiming to correspond to the dynamics of the events, but with the events being broken down into data captured at particular time points of the event. For example, a traffic jam may be modeled as a process, but the cartographic representation usually relies on a series of "snapshots" created to show the status of the traffic jam at different points in time. The same goes for natural events: the spread of a wild fire modeled as an event may use the same technique to show in an animation how the wild fire spreads.

In other words, while the underlying concepts and geographic representations can take both patterns and processes into account, the cartographic representations usually only show an animation of "snapshots" prepared to show the event's development.

January 25, 1983   February 8, 1991   April 29, 1997

September 4, 2000

FIGURE 13.1. Growth of Austin, Texas, showing urbanized areas in 1983, 1991, 1997, and 2000.

From *http://tx.usgs.gov/geography/austgrth_large.htm*

SIMPLE/COMPLEX

GI analysis can range from simple to complex. *Simple* refers to activities involving interpretation or comparisons. *Complex* refers to geostatistical analysis and process modeling. In between simple and complex types of analysis lies a vast range of analysis types and transformations. For example, a simple geographic information analysis may consist of merely comparing two data sets of the same area collected at different times, combining a forest-type data set with a soils characteristics data set, or using buffer operations to determine the service area of a proposed bus route change. Complex analyses can use Monte Carlo simulations to assess which distribution of soil pH values most likely matches the stochastic distribution, use fuzzy-set theories to assess the inaccuracy of boundaries around vegetation types, or rely on variance calculations to help determine the reliability of field data samples.

Many or even most geographic information analyses lie in between these two poles. They will rely on some transformations and an interpretation or comparison. For example, creating a buffer around a factory which is submitting an expansion plan for the facilities is the transformation necessary to combine this buffer with the positions of buildings whose owners need to be informed about the permit application. Many transformations can be linked to a single interpretation or comparison, or each transformation may involve hundreds of data sets that are combined for a single interpretation of how multiple factors influence each other.

ACCURACY/RELIABILITY

Accuracy refers to the degree of correspondence between data and the actual thing or event. *Reliability* indicates how consistent the data is for certain types of applications. Usually high accuracy means high reliability, but in several circumstances the opposite may be the case.

First, data may be highly accurate, but because of the time between its collection and its analysis, it may no longer be reliable. Aerial photos and satellite imagery can easily become dated and will vary greatly from the actual situation. Second, reliable data may have a low accuracy. For example, data showing the major roads in the United States may be useful for reliably determining how to go from Boston to San Diego, but not accurate enough for determining how far a recycling center in San Diego is located from a highway.

The balance between accuracy and reliability is often a financial issue. Because of the high costs of data collection, often limitations for both accuracy and reliability are acceptable. The key points about this balances is first to clearly describe the data's date of collection and concerns about possible discrepancies to the actual situation and, second, to take these discrepancies into account when analyzing the data.

Basic Geographic Information Analysis Types and Applications

Solely as an overview of the many types of geographic information analysis and applications, this section gives some insight into pragmatic issues and applications of geographic information analysis. These geographic information analysis operations only partially correspond to GIS operations and commands. Depending on the software's analytical capabilities, the analysis in GIS may involve a single command or many commands. Because of the endless permutations of geographic information analysis operations, a direct match to any single GIS software's set of operations and commands only makes sense for specific applications and domains.

Query

People who have worked with maps may think of the geographic information query in terms of interactive maps. In its simplest form, geographic information querying involves both spatial and attribute aspects allowing for a person to select a feature and find out its attributes. In many applications, the programming of the interface can make this a very helpful feature to find out the name of lakes, cities, clinics, or restaurants. Even for professional geographic information analysis, geographic information querying is often important in determining attributes of individual features. Related to the

TABLE 13.1. GI Analysis Types and Related GIS Operations

Type of GI Analysis Operation	Related GIS Operation
Query	Spatial query and attribute query
Combination	Overlay
Distance transformation	Buffer
Neighboring	Connectivity, adjacency, visibility
Rating	Ranking, weighting
Multivariate analysis	Linear factor combinations
Geostatistics	Monte Carlo simulations
	Fuzzy sets
	Variance

Note that GIS operations may not correspond to actual GIS commands.

basic "identify" operation are operations for determining the characteristics of features based on their geographical relationship to other features, particular positions, or areas. Another type of query operation makes it possible to select features based on their attribute values. This type of query is a significant analytical tool for choosing features based on combinations of attributes. It is often used after geographic information is combined to identify particular combinations of attributes.

Combination

One of the most used geographic information analysis types is widely known by the associated GIS operation: overlay. Overlay in its simplest form involves joining the vector or raster data from two data sets including the attributes. The people who introduced overlay to GIS thought of this type of geographic information analysis in terms of looking at interactions between two or more transparent maps. The simplicity of this approach was very attractive, as was its chief promoter's (Ian McHarg) emphasis on the overlay operation as a method to incorporate environmental and social concerns along with engineering perspectives in the planning of highways and other large construction projects.

Many people refer to overlay as "integration," but each GIS overlay operation may or may not integrate. As McHarg, Dangermond, and many others from the first generation of GIS developers point out, integrative analysis based on overlay requires the interpretation of the overlay results. Simply combining two (or more) data sets through overlay will only rarely integrate geographically.

The combination geographic information analysis operation is prolific due to its ability to take geographic information from different sources. As long as the geographic information uses the same coordinate system, it is possible to combine the GI, although most times additional query geographic information analysis operations are required before or after the combination operation.

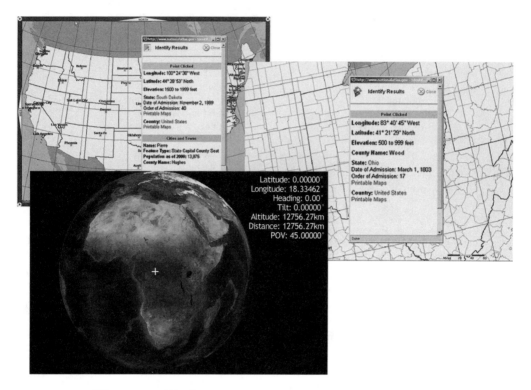

FIGURE 13.2. Different types of identify operations.

Distance Transformation

The second most used geographic information analysis type is usually known by its GIS operation equivalent: buffer. The distance transformation transforms a feature or area of raster cells into an area based on given distances. One single distance can be used—for example, 100 feet from the well—or multiple distances—for example 50 m, 150 m, 250 m, and 500 m from the roads—for the transformation. The distance transformation geographic information analysis is often used to show the geographic extent of events (e.g., noise from traffic, leaking of oil tanks into the ground) as a thing. In these uses the distances correspond to model or assumed values regarding the processes underlying the events. This ability to transform from process to pattern is perhaps the single most important reason for the significance of this geographic information analysis type.

Neighborhood

Though related to distance transformations, neighboring is focused more on establishing what and how features are geographically related. Neighborhood geographic information analysis usually either focuses on using topol-

FIGURE 13.3. The combined values of the integrated terrain units must be analyzed to be meaningful.
Source: Dangermond (1979). Adapted by permission.

ogy, raster cell neighborhoods, or TIN relationships to analyze geographic relations. For geographic information analysis of transportation networks, these geographic information analysis operations are critical to checking and establishing different types of connectivity. In environmental applications, neighborhood geographic information analysis is used for a variety of applications including modeling soil erosion, establishing water runoff patterns, and determining viewsheds.

Rating

Often following or preceding other geographic information analysis operations, rating is used to ordinally rank features based on combinations of attributes or combinations of attributes and locations. It is also often the operation that is a key part of weighting various attributes for decision making. This process is prone to distortions, requiring that great care be taken in determining the weights.

FIGURE 13.4. Distances can be transformed between raster and vector formats. These transformations can lead to distinct differences in the corresponding areas.

FIGURE 13.5. Examples of neighborhood operations using raster GIS.

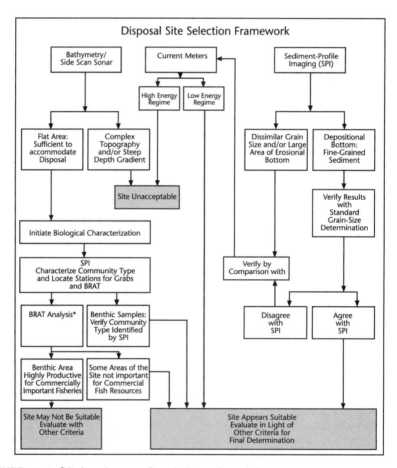

FIGURE 13.6. Criteria and process for selecting a disposal site.

Multivariate Analysis

To avoid some of the problems of weighting, multivariate analysis is a geographic information analysis operation that facilitates more flexible permutations of attribute combinations and relationships between attributes and geographic positions. The variables are often combined in linear equations that simply provide for the additive combination of attributes and relationships, but multidimensional equations provide ways to consider attributes and relationships in more complex situations.

Geostatistics

The most complex geographic information analysis operations, and, for many reasons, the most important geographic information analysis operations, are geostatistical geographic information analysis operations (covered in Chapter 14). Considering even just three of these rich operations can point out the importance of geostatistics especially for applications that need to go beyond geographic modeling of things as patterns and transform-

FIGURE 13.7. Site selection for disposal of marine sediments. The characterizations used in this decision making are used to determine the relative importance of habitats for commercial fishing and suitability for depositing marine sediments.

From *www.csc.noaa.gov/benthic/mapping/applying/pdf/bmdredge.pdf*

ing events into patterns. *Monte Carlo simulations* offer the means to create data that stochastically offers a reliable estimation of the geographic differentiation of continuous characteristics—for example, soil pH, species densities, or transportation costs. *Fuzzy-set theory* is the foundation for a number of geographic information analysis operations that consider the variability of natural phenomena boundaries and the inaccuracies of data collection. *Variance* measures the difference between repeated measures of the same properties. It is important in assessing the impacts of different geographic aggregation units on the accuracy of data.

GIS in a Nutshell

With many text books offering varying introductions to GIS, the purpose of this section is only to show GIS's relationship to geographic information analysis operations and the role of GIS in representation and communication. Consider two definitions followed by a third that brings aspects of them together.

First Definition

A system of hardware, software, data, people, organizations and institutional arrangements for collecting, storing, analyzing, and disseminating information about areas of the earth (Duecker & Kjerne, 1989).

This commonly cited definition focuses on important parts of a system that consists of six components and is used for four generic purposes. It can be used to describe any organized use of geographic information. GIS definitely involves these components and is used in these four ways. If we consider the terms very broadly—for example, hardware includes notecards and software includes alphabetical filing systems—then even an address list could be considered to be a GIS. In this definition, GIS is used as a single system for all elements of processing GI, beginning with collecting and ending with making maps and other types of information.

What about the geographic representation and cartographic representation? Duecker's and Kjerne's definition provides insight into what GIS can be used for. What about how, by whom, and for whom? If you know what the GIS is being used for, then this definition is practical because it lends great flexibility for actually using a GIS in many different ways. It also may be too vague in how it explains the relationships between the components and generic purposes. Are all components equally involved in storage? This seems to be a naïve question, but with this definition standing on its own, as a definition should, you really couldn't tell. This definition is handy, but what most people understand when they rely on this definition is just the surface of GIS; the geographic representation and cartographic representation are missing.

Second Definition

Organized activity by which people measure and represent geographic phenomena then transform these representations into other forms while interacting with social structures (Chrisman, 1999).

Nicholas Chrisman developed this definition as an attempt to address the open questions about Duecker's and Kjerne's definition (and many others—see chapter readings—that express the same main concepts as their definition). Chrisman's definition focuses on the activities of measuring and representing in the context of social structures. This is the short form of a more involved conceptual model of GIS which consists of a nested set of rings and interactions between the rings.

Chrisman's "shell model" of GIS is broad and inclusive. In its focus on activities, it points to the importance of knowing what any particular GIS is used for. Operations (common GIS processes) and transformations (processes that change the measurement framework) are the emphasis in Chrisman's discussion of how GIS is used. In the redrawn and modified version of the "shell model" figure, the emphasis is placed on activities that are essential to the successful development and use of any GIS. First, data quality involves verifying measurements and geographic representations in comparison to the corresponding things and events found in the world. Second, the operations and transformations used in any GIS need to undergo an evaluation of each and every use of the GIS. Finally, conventions originating in society, culture, and institutions require consideration to assure that the data accurately correspond to the things and events and the operations and transformations take into account the goals of the GIS.

Third Definition

System of computer hardware, software, and procedures designed to support the compiling, storing, retrieving, analyzing, and display of spatially ref-

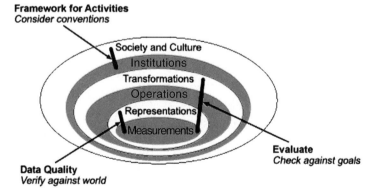

FIGURE 13.8. "Shell model" showing elements of GIS and interactions

Source: Chrisman (1999). Adapted by permission.

erenced data for addressing planning and management problems. In addition to these technical components, a complete GIS must also include a focus on people, organizations, and standards (*http://www.extension.umn.edu/ distribution/naturalresources/components/DD6097ag.html*).

This definition merges parts of the first and second definitions specifying purposes and the importance of "people, organizations, and standards." It starts out with the components of Duecker's and Kjerne's definition, adding more specific purposes and rationales and notes the importance of non-technical elements.

Definitions like these provide more substance for people unfamiliar with GIS, but very interested in using GIS in their organizations. In this definition one of the purposes is "management problems." It also broadens the consideration: this is not just a matter of technology. The people working on the problems, the organizations, and the standards used in developing the GIS must also be considered.

One thing these definitions fail to cover is that "layers" are a common term for describing the organization of data in a GIS. The "layer" organization is very significant. It allows for the storage of multiple geographic representations of the things and events in the same GIS. If they use the same coordinate system, they can even be combined. This concept underlies the MPLIS and SDI concepts, even if the combination of two "layers" can lead to many problems because of differences in geographic representation, often detected as problems with accuracy. "Layers" are the visual concept; in databases the data can be stored in various formats. The term "layers" helps people imaging the potential relationships and combinations of different geographic representations.

Example GIS Applications

Out of the plethora of GIS applications, this section presents a few vignettes that are illustrate the roles of geographic representation and cartographic representation and the significance of accuracy, goals, and conventions in developing GIS applications.

Environment and Conservation Applications

LANDSLIDE ANALYSIS

In Japan a GIS application was developed to store and manage three-dimensional surface and subsurface data in Akita, Yamagata, and Kanagaw prefectures. The application allows users to construct flexible analysis and examine the effects of preventative structures and plantings, and it can actually simulate landslide movements. Its geographic representation takes various hydrogeological characteristics into account, which can be linked to topographic maps, technical drawings, and orthorectified photos. Organized into layers, the different elements of the geographic representation can be analyzed based on geometric overlaps. Because of the three-dimensional nature of landslides, the cartographic representation supports the use of multiple viewpoints and user-defined visualizations of cut-lines that help users to understand and examine the geological structure. The analyses and visualizations need to be very accurate to meet the goals of users and established conventions of geohazard analysis.

SEA CLIFF EROSION

The municipality of Isla Vista, California, near Santa Barbara, has a very high population of students. Apartments overlooking the Pacific Ocean are in high demand, but ongoing erosion has led to several apartments being condemned. Students from the University of California, Santa Barbara, were involved in creating a GIS that mapped details of the cliff edge and the cliff base that could be used in the county GIS. Existing data from the county GIS was first analyzed and then students went out to collect additional data using GPS and video cameras. To assure that the data was accurate, the cliff data on erosion activity, storm drains, vegetation, and beach access points was collected four times. Using county data, students also analyzed the rate of coastal erosion since 1972. The changes, on average just less than 1 foot of cliff erosion per year, were used to make a prediction of the coastal cliff changes through 2055. The cartographic representations included maps, animations, and interviews. The high accuracy of the data points to the importance of having robust scientific data to fulfill conventions for data that will be used by the municipality and county in making decisions.

FIGURE 13.9. Section of map produced with a GIS-based analysis of landslide hazards on Puget Sound.

From *http://pubs.usgs.gov/mf/2000/mf-2346/mf-2346so.pdf*

Urban Applications

IMPROVING LAND ADMINISTRATION

In many countries around the world, cadastral records, planning documents, and planning maps needed for land administration are spread among various agencies. Collecting and comparing documents can require a great deal of time, and problems and conflicts often arise. Many GIS applications have been developed to assist governments with land administration. These applications must take the existing geographic representations into account, but also determine what are common agreements and disagreements between the government agencies. By developing a geographic representation for the GIS that helps facilitate interactions and flags possible problems for resolution, the GIS can greatly improve government land administration. Developing cartographic representations that accurately indicate different planning zones and ownerships, and that help administrators understand the problems, are critical to the success of these applications. Through this support and using information technology to improve the accuracy and speed of processing, land administration can make important decisions much quicker.

MODELING URBAN GROWTH

The rapid increase in urban populations around the world (more than 50% of the world's population now live in cities) leads to a number of health, social, and environmental problems. To help administrators and politicians develop a better understanding of this growth, research projects modeling the process of urban growth have helped predict the future growth of areas. These studies require very careful consideration of geographic representation issues. For instance, people will tend to migrate to areas with good transportation to employment possibilities, but these areas often have the highest rents. Many people will look for nearby alternatives that help them save on housing costs, but are still close to transportation. The development of an area can accumulate and lead to very fast growth as more people choose a place to live, more transportation is to be provided, more people come, and housing costs increase. After a certain point people will move to new nearby locations, starting the cycle over again in a new place. The geographic representation of these interactions requires detailed modeling of the numerous political, social, and economic factors that influence the process of urban growth. The detail of the geographic representation is also important for assessing the accuracy of the model and assuring that the goals of the modeling are met. The cartographic representation of these models needs to take conventions for mapping an area into account to assure that people can understand the results of a model. See Plate 12 for an example of a field model.

Summary

Geographic information analysis improves spatial decision making and communication. It takes many forms, ranging from the comparison of data sets to dynamic models of interactions. The choices made in analysis reflect the same conventions involved in any type of geographic representation or cartographic representation. Three issues are especially significant for geographic information analysis: (1) choosing patterns or processes, (2) applying simple or complex analysis techniques, and (3) determining the appropriate balance between accuracy and reliability. Basic geographic information analysis techniques include analysis, combination (overlays), and distance transformation (buffers). More complicated techniques include neighborhood analysis and rating. Multivariate analysis is a more complex form and geostatistics is the most complex form of analysis.

Geographic information analysis relies on GIS, which is also used for cartographic presentation. GIS is used for every kind of geographic information analysis and has become as significant as the microscope in changing how people analyze the world.

Review Questions

1. What should GI analysis consider so as to communicate the desired intent?
2. What is the difference between patterns and processes?
3. Why distinguish between simple and complex types of analysis?
4. Which GI analysis type is more common, buffers or overlay?
5. Is there a trade-off between accuracy and reliability?
6. How is Chrisman's definition of GI different from the "input–process–output" definition?
7. What are the elements of Chrisman's GIS definition?
8. How does the limited ability of GIS to consider process constrain considerations of events?
9. What is the difference between a translation and a transformation?
10. What is a common application of buffers?

Answers

1. What should GI analysis consider so as to communicate the desired intent?

   GI analyses should consider the history, culture, and purposes along with issues of geographic representation and cartographic representation.

2. What is the difference between patterns and processes?

   Patterns are geographic representations that portray a static geographic situation or a snapshot of an event. *Processes* represent the geographic interactions dynamically.

3. Why distinguish between simple and complex types of analysis?

This distinction helps one to grapple with analysis applications that make more of the choices underlying the analysis visible.

4. Which GI analysis type is more common, buffers or overlay?

Neither. Both are of great importance generally. Specific disciplines or applications may use one or the other more often, but across the board both are very important.

5. Is there a trade-off between accuracy and reliability?

Generally not, but in cases where positional and temporal accuracy are both involved, there may be a trade-off between lower positional accuracy and greater temporal accuracy or vice-versa.

6. How is Chrisman's definition of GI different from the "input–process–output" definition?

Chrisman's definition accounts for different interactions and context issues.

7. What are the elements of Chrisman's GIS definition?

The four "activities" in Chrisman's GIS definition are making measurements of geographic phenomena and processes, making representations of what was measured, performing further operations, and making transformations to other representational systems.

8. How does the limited ability of GIS to consider process constrain considerations of events?

Events in most GIS need to be geographically represented as things. A set of things ordered by the time of their observation can be used to create an animation, but this dynamic visualization does not necessarily correspond to the actual process.

9. What is the difference between a translation and a transformation?

Translations are done by humans, usually working with geographic information on computers, but sometimes working with only maps and other printed material. Transformations are done by GIS software to produce new geographic information from existing geographic information, for example, a buffer to generate the extent of an animal's biotope from the site of its nesting.

10. What is a common application of buffers?

Buffers are commonly used to create a zone that corresponds to the effects of a process. For example, buffers offer a crude way to represent the spread of noise from vehicles, airplanes taking off and landing, or pollution emissions from a smokestack or outlet pipe.

Chapter Readings

Chrisman, N. R. (1999). What Does "GIS" mean? *Transactions in GIS*, 3(2), 175–186.

Dangermond, J. (1979). A Case Study of the Zulia Regional Planning Study,

Describing Work Completed. In G. Dutton (Ed.), *Urban, Regional and State Applications* (Vol. 3, pp. 35–62). Cambridge, MA: Harvard University Press.

Goodchild, M. F. (1978). Statistical Aspects of the Polygon Overlay Problem. In *Harvard Papers on GIS: First International Advanced Study Symposium on Topological Data Structures for Geographical Information Systems*. Cambridge, MA: Harvard University Press.

Haklay, M. (2004). Map Calculus in GIS: A Proposal and Demonstration. *International Journal of Geographical Information Science, 18*(1), 107–125.

MacDougall, E. B. (1975). The Accuracy of Map Overlays. *Landscape Planning, 2*, 25–30.

McHarg, I. (1969). *Design with Nature*. New York: Natural History Press.

O'Sullivan, D., & D. J. Unwin. (2003). *Geographic Information Analysis*. New York: Wiley.

Tomlin, C. D. (1990). *Geographic Information Systems and Cartographic Modeling*. Englewood Cliffs, NJ: Prentice Hall.

Veregin, H. (1995). Developing and Testing of an Error Propagation Model for GIS Overlay Operations. *International Journal of Geographical Information Systems, 9*(6), 595–619.

Web Resources

For a broad overview of GIS analytical use (with an emphasis on caving), see *http://rockyweb.cr.usgs.gov/public/outreach/articles/nss_gis_article.pdf*

Examples of the application of GIS-based GAP analysis are available online at *www.gap.uidaho.edu/applications/applications.htm*

Applications from around the world showing uses of GIS and remote sensing in urban analysis are available online at *http://web.mit.edu/urbanupgrading/upgrading/case-examples/index.html*

Archaeological application involving GIS-based analysis are available online at *www.informatics.org/france/france.html*

Exercise

1. *Describing and Evaluating a GIS Application*

Concepts

GIS involves the organization of many different aspects into a system. These aspects are interdependent. In this exercise, you will discuss the aspects of a GIS in terms of the definitions from Chapter 13.

Objectives

Write an essay that systematically describes the components, issues, and activities of a GIS application.

Activities

Using Chrisman's definition, organize the outline for an essay that includes the components, issues, and activities described in the text. Make sure to address the accuracy, conventions, and goals of the application you choose.

Chapter 14

Geostatistics

Geostatistics involves both the most complex and the most important GI analysis operations in part because of the broad uses of geostatistical analyses operations, but perhaps even more due to the underlying power of the mathematical analysis of GI. For many people geostatistics is far easier to grasp than abstract mathematics because its mathematics are tied to actual things and events.

This chapter provides an introduction to the concepts of geostatistics. First, we examine the concepts by themselves, followed by discussion of some applications. Geostatistical applications run the gamut of statistics, applying techniques and concepts from classical probabilistic statistics to Bayesian-based statistical analysis. In many environments these applications are only the first cut of more detailed analysis required for assessing geographic patterns, processes, and relationships.

Patterns Indicate Processes

In geostatistics, patterns express evidence of spatial processes. At first you may think of pattern in a visual sense—for example, a map showing temperatures across the United States. But, in geostatistics, the visible pattern points to underlying relationships, which can be expressed mathematically. The geographic distribution of temperatures helps illustrate this point. Following fundamental physical laws, the temperatures of any body will tend toward equilibrium that is, temperatures that start out very different will rise or fall toward an equilibrium point, somewhere around the average. But clearly weather, with cold fronts, warm fronts, winds, human influences, jet streams, clouds, and many other characteristics, is dynamic. Temperature is just one indicator at a particular place of the weather. And the temperatures measured at various places in an area at a particular point in time form a pattern that indicates relationships between fronts, jet-stream, and so on. By considering temperature measures made at other points in time in the area, the comparison of different temperature measures leads to some understanding of the underlying processes.

These comparisons can be made visually, but are then plagued by uncertainty. Scale is an important factor: if we consider temperature differences in the entire United States, then it will be very difficult to tell clearly what the differences in temperatures are between Philadelphia and Baltimore. Visual symbols, especially color, may be easily misinterpreted. The comparisons will also run into difficulties when a person tries to compare temperature changes over a year, or even over a season, if the detail is significant, because of the large number of temperature maps. A person might be able to interpret differences between five maps of the United States, but comparing 500 temperature maps would be impossible for almost anybody.

Geostatistics takes a mathematical approach and works directly with the underlying measurements. By working directly with the measurements, one can examine the relationships between places and the processes in more detail regardless of scale, regardless of different visual interpretations of color, and regardless of the number of maps. In other words, even if the statistical operations are simple comparisons, mathematics helps one to get a grasp on the complexity of mapping far more readily than visual techniques. Additionally, because the measurements are being used directly, they can be used for other analysis operations that can aid one in understanding the accuracy and validity of the measurements and provide insights into climatic processes both at the places where the measurements are made and the places in between.

Patterns indicate characteristics or phases of processes, but all relationships between patterns and processes are very complex. Other factors may need to be considered for more accurate insights into weather processes. While temperature is important, it is only one indicator of weather. It also varies considerably during the day, based on exposure to winds and the sun, according to locations, and so on. Measurements of these aspects should be considered with temperature to gain a better understanding of the processes.

You also need to be aware of the complexity of establishing, creating, and validating measurements. When measurements are collected on the environmental characteristics of things and events, they are usually collected at distinct places. How these places are determined is a crucial question for assessing the reliability and validity of the measures. The sampling distribution must be carefully decided on based on an analysis of known properties and processes, or through a simulation of the patterns and process to be studied. The types of instruments and the types of measurements are another important point to consider. If different instruments are used in different places, the measurements may be significantly different. If the exact differences can be ascertained, it may be possible to transform the measurements to a common reference system. For example, measurements of temperature in Fahrenheit and Celsius need to be converted to one system

FIGURE 14.1. Example of environmental data sampling showing how data is combined in geostatistical analysis.

From *www.nwer.sandia.gov/sample/ftp/tutorial.pdf*

before they can be compared. Both the establishment and the creation of measurements have significant impacts on the validation of measurements. Validation is an important separate part of data collection, necessary to assure the internal validity of measurements made at the same place over time and the external validity of the measurement to other places, other instruments, and other types of observing and recording the same types of measurements.

Spatial Autocorrelation

Geostatistics is powerful, but easily prone to great errors through simple misrepresentations or because of great differences between most types of statistical data. The most significant problems are that geographic data sets are not random, nor samples. They are spatially autocorrelated. As Waldo Tobler notes, "I invoke the first law of geography: everything is related to everything else, but near things are more related than distant things" (Tobler, 1970). A basic principle of "classical" statistics is that data are randomly distributed—for example, the ages of people in a city have nothing to do with the city ("classical" distinguishes the statistical concepts and techniques discussed here from Bayesian statistics). Obviously, geographical data are affected, sometimes even determined, by their location. Amphibians flourish near water bodies, schools are located near where children live, stores are accessible to shoppers, and traffic jams occur on busy roads. The concept of Tobler's first law may seem blatantly obvious to geographically minded people, yet it is completely at odds with the principle of "classical" statistics. ("Classical" distinguishes these from Bayesian statistics.) Geographical things and events can sometimes be more complicated. For example, infectious diseases can spread by plane travel across oceans. Geographers call this "jumping scale" because the infection, which normally spreads at a local scale, "jumps" globally and becomes active at the local scale again.

Following the concept of spatial autocorrelation, it is easy to grasp that samples—for example, measurements of temperature or anything else—cannot reflect a completely uninfluenced area, where the measured characteristics follow a random distribution. For example, let's assume that most frogs prefer a partly wet habitat. If a researcher is determining the impact of household pesticide use on the local frog population, it makes the most sense to collect frogs near water bodies in low-lying wet areas. However, that

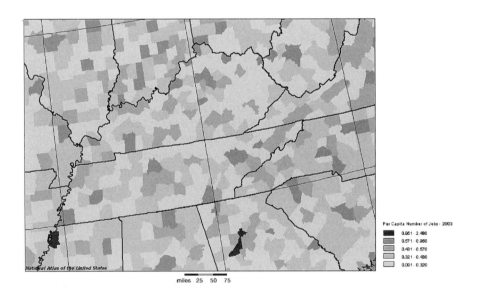

FIGURE 14.2. *nationalatlas.gov* offers an interactive mapping application—here, the per capita number of jobs for a portion of the country is shown.

From *www.nationalatlas.gov/natlas/Natlasstart.asp*

may or may not be the places where people live and apply pesticides. The relationship of residential location and wetlands needs to be considered in collecting the frogs. The basic principle applies to any type of geostatistical data collection: all sampling and collection of data needs to take account of the underlying processes and factors that influence the processes. The complexity of relating sampling and data collection to things and events makes geostatistics very complex, yet, when properly and reliably done, very reliable.

The Ecological Fallacy

A significant mistake easily made when working with geostatics is based on what is known as the "ecological fallacy." The *ecological fallacy* is the assumption that the statistical relationship observed at one level of aggregation holds at a more detailed level. A well-known example of this occurs when people look at statistics of election results. In most U.S. states in the 2004 election, the majority of voters voted for George W. Bush, but most people in the cities voted for John Kerry—if you only counted the votes from a city and assumed it applies for the state, you would be in error. Take another example: There may be a strong relationship between the number of zebra mussels in lakes and the number of recreational boaters in a state, but other local and regional factors in a state may be more significant in explaining the relationship. The strong statistical significance at the state level may be hid-

FIGURE 14.3. Cartograms show how other geographical characteristics influence mapped attributes, issues that geostatistics can take into account.

From *www-personal.umich.edu/~mejn/cartograms/*

ing the factors that actually lead to large numbers of zebra mussels. Something completely different may be causing the high number of zebra mussels. Perhaps it is the size of lake, the number of boats using multiple lakes, or the number of freeze-over days. In every case of geostatistics, the relationship between the aggregation units and the things and events being studied must be carefully examined.

Modifiable Areal Unit Problem

The modifiable areal unit problem (MAUP) is a special instance of the ecological fallacy that results when data collected at a more detailed level of aggregation—for example, census blocks, counties, or biotopes—are aggregated to less-detailed levels of aggregation—for example, census tracts, states, or watersheds. The aggregation units may be arbitrary to the things and events being studied, but the aggregation units used in collecting or collating the data will affect the statistics that use this data. The consequences of the MAUP can be significant. Although the assignment of counties to states is political, statistics show that a switch of one northern Florida county to Georgia or Alabama would have produced a different outcome in the 2000 U.S. presidential elections.

Terrain Analysis

Terrain analysis is an important application domain of geostatistics for a number of disciplines and professions. Civil engineers rely on terrain analysis when planning the construction of large structures, cellphone companies use terrain analysis to plan the siting of antennas, city planners rely on terrain analysis to assess the impacts of new buildings on the landscape, and the military uses terrain analysis for planning and preparing missions. The list could go on.

Because of the breadth of applications and the number of operations and variables, the role of geostatistics in terrain analysis is hard to define. To begin, you can distinguish types of terrain analysis by the role of visual interpretation in the application. Some applications begin with a visual analysis of field data, maps, aerial photographs, remote sensing images, or various combinations of these materials. Other applications may start out with geostatistical analysis or rely on geostatistics to analyze the materials discovered during a project.

For example, many archaeologists use terrain analysis by beginning with a visual inspection of aerial photographs or remote sensing data to see if traces of previous habitation or structures are visible. After comparing these materials with previous archaeological projects and documents, the researchers often look to geostatistical techniques as a means of examining multiple sources of data and evaluating characteristics in the data sources for relationships that can help with understanding the previous cultures in an area. A common application of geostatistics begins with analyzing previous habita-

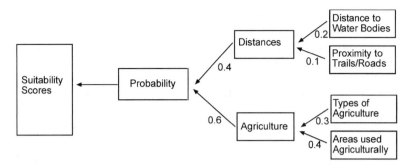

FIGURE 14.4. Weighted factors are often used to determine suitability. The weights adjust the factors for subjective values.

tion patterns to establish geographical relationships. Among these relationships are distances of settlements to water bodies, trails and roads, and areas with different types of agriculture. These relationships can be quantified as measured distances and used to establish areas of potential settlements, agriculture, and habitation. These areas can be combined by using an overlay operation or geostatistically analyzed to assess the combined potential that any particular area was used by the past culture. This type of application is helpful in planning potential archeological digs. It can be extended and detailed in a variety of ways.

The use of geostatistics in archeological analysis such as this example may involve a simple factor analysis of the geographical relationships. A *factor analysis* is an explanatory analysis technique that lacks an assumption that the factors are independent. The factor analysis is therefore prone to substantial variability arising in the determination of factors and the assignment of weights (see Figure 14.5). Why, for example, does the distance to water bodies have a weight of 0.2 while the proximity to trails and roads only has a weight of 0.1? The sense of these weights is not transparent, but takes into account the specialized knowledge of the archeologist. The explanation may help, but if the factors and weights are badly chosen, they may lead to misleading results.

Types of Terrain Analysis

Another way to distinguish different types of terrain analysis is to distinguish between two-dimensional and three-dimensional terrain analysis. While there is a fuzzy boundary between the two and this division means separating the visual presentation of results from the analysis, the distinction helps one to get a basic grasp on fundamental types of terrain analysis.

Two-dimensional terrain analysis, or 2-D, usually relies on the use of raster data to analyze relationships based on location. A common 2-D raster terrain analysis is visibility analysis called "viewshed analysis." This operation assesses the area of cells that can be seen from an origin, a raster cell, by comparing the elevation of the neighboring cell with the origin and going outward to other cells if the cell in question is at an elevation equal to or below the elevation of the origin cell. This type of terrain analysis is important in environmental mediation, planning, and the location of transmission towers. It also can be used as part of more complex terrain analysis—for example, considering the impact of large-scale landscape changes on snowmelt processes.

More complex raster-based terrain analysis is used for countless environmental applications. Watershed processes, erosion, and sediment yield analyses often use terrain analysis as part of more comprehensive analysis that takes account of dynamic processes that are ill-suited for factor analysis. A sediment yield analysis, for example, can consider soil types, vegetation, geology, maintenance practices, water absorption capacity, and numerous other factors as part of a dynamic principal component analysis that accounts for the amount of precipitation and duration as key variables. Principal component analysis aids the simplification of complex statistical relationships through the identification of independent and uncorrelated variables.

Vector-based terrain analysis is used less frequently because the GI representation of vectors in commercial GIS produces sharp boundaries that generally do not reflect the field nature of environmental things and events. It does find many applications in detailed engineering work—for example, in

FIGURE 14.5. Archaeological least-cost model of the possible path of a Roman road in England. Slope is considered to be the primary determinant of the possible path of the road.

Courtesy of Jason Menard.

the calculation of cuts and fills in construction projects, or the determination of elevations.

Three-dimensional terrain analysis (3-D) is most commonly performed with TIN-based terrain analysis or, less frequently, with 3-D rasters, called "voxels." TIN is used in analysis of surficial processes—for example, determining erosion or assessing watershed run-off. TIN-based terrain analysis often uses a digital elevation model (DEM) and water body or road network data. The geostatistics involved in this are commonly linear equations, but they can involve principle component analysis, fractals, and shortest path analysis. Voxels are used in specialized applications to analyze relationships in the atmosphere or in the ground, including the diffusion of aerosols from factories and the pollution of groundwater.

Chi-Square Analysis

Chi-square analysis is a straightforward statistical technique that can be used to evaluate the validity of a hypothesis. It is used to examine the relationship between two variables in a cause-and-effect relationship. Because it is reasonably straightforward, it is widely used for exploring a number of questions. For example, does higher soil pH lead to less healthy plants? Does proximity to bus stops lead to an increase of people using public transit? How strong is the relationship between crop types and water resources? It may need to be followed up by more exacting geostatistical study, but chi-square analysis often serves as a key starting point for testing and validating questions. Chi-square analysis compares an idealized random distribution with existing or projected distributions. The random distribution of variables for chi-square analysis that measure characteristics of things or events is called the "normal distribution," or what one would expect if there is no relationship between the variables. The existing distribution is called the "expected distribution." We can go through an example, step-by-step, to get a better ideal of how chi-square analysis works.

1. Create an Observed Frequency Table and Examine Relationships

The first step is to organize the data into a contingency table where the rows indicate one variable of the independent variable (considered the "causal" factor of the relationship) and the dependent variable (the "effect" from the independent variable). For this example, we will consider how elevation effects snowfall. Our hypothesis is that more snow falls at higher elevations. We have the data from 133 observations at elevations between 500 and 4,500 m and the yearly average amount of snow, which ranges from 0 to 534 cm. In Table 14.1, the observed frequencies table, the rows indicate elevations, classified into three groups, and the columns indicate the snowfall averages, classified into three ordinal categories. The individual cells give the number

TABLE 14.1. Observed Frequencies

| Elevation | Yearly Snowfall | | | |
	Low (0–20 cm)	Medium (20–60 cm)	High (60 + cm)	Totals
500–1000 m	41	19	10	70
1000–2000 m	22	15	6	43
2000 + m	2	8	10	20
	65	42	26	133

of cases that fit into the combination of elevation and snow—for example, 10 observations at elevations over 2,000 m have, on average, more than 60 cm of snow yearly. Looking over all the cells, one sees a marked tendency in the observations that suggest higher elevations receive more snow than lower elevations do. The chi-square statistic is used to test the relationship between the two qualitative variables.

2. Formulate a Test Statement

Each distribution accounts for both independent and dependent variables. The test statement expresses the relationship we think we see in the data in a more specific manner. The established approach for creating this *null hypothesis* is that it states that the variables are not associated. If the chi-square statistic disproves the null hypothesis, then the opposite is proven, namely, that more snow falls at higher elevation. In statistical terminology, H_0 refers to the null hypothesis; H_1 refers to the alternative hypothesis.

We use the chi-square statistic to determine the difference between the actual observations and what we would expect if the observations followed our null hypothesis. We need to create a second table based on the assumption that the null hypothesis is correct and then compare the two tables. The values for the second table, called the "expected frequencies," are calculated by using the row and column totals. First, calculate the expected probability that the snowfall is low by dividing the total number of observations of low snowfall by the total number of observations (round the results to three significant digits):

TABLE 14.2. Expected Frequencies

| Elevation | Yearly Snowfall | | | |
	Low (0–20 cm)	Medium (20–60 cm)	High (60+ cm)	Totals
500–1000 m	34.209	22.107	13.648	70
1000–2000 m	21.006	13.575	8.379	43
2000+ m	9.709	6.251	3.79	20
	65	42	26	133

TABLE 14.3. (Observed – Expected)²

	Yearly Snowfall		
Elevation	Low (0–20 cm)	Medium (20–60 cm)	High (60+ cm)
500–1000 m	46.118	9.653	13.308
1000–2000 m	0.988	2.031	5.660
2000+ m	59.429	3.059	38.564

Expected probability (low snowfall) = 65/133 = 0.489

Second, calculate the expected probability that the observations are between 500–1,000 m:

Expected probability (500–1,000 m elevation) = 70/133 = 0.526

Since the null hypothesis assumes they are independent variables, calculate the combined expected probability by multiplying the two expected probabilities together:

Combined expected probability = 0.489 × 0.526 = 0.162

Then multiply the total number of observations (133) by the combined expected probability to determine the expected frequency of low snowfall at 500–1,000 m elevation:

Expected frequency = 133 × 0.257 = 34.209

Repeat these four steps for each relationship (the other eight cells) between snowfall and elevation to complete the expected frequencies table.

Remember that the expected counts in each table square are not the actual observations, but are based on the assumption that there is no relationship between the two variables. If we even visually compare the two tables, we can see differences suggesting that the idea that more snow falls at higher elevations is probably right and that the null hypothesis will be disproven by the chi-square statistic. To calculate the chi-square statistic and

TABLE 14.4. (Observed – Expected)²/Expected

	Yearly Snowfall			
Elevation	Low (0–20 cm)	Medium (20–60 cm)	High (60+ cm)	Totals
500–1000 m	1.348	2.086	3.379	6.813
1000–2000 m	2.195	3.397	5.504	11.097
2000+ m	4.750	7.378	12.168	24.296
Total	8.294	12.861	21.051	42.206

have more certainty than our visual inspection, we need to first square the differences between the observed and the expected counts:

Squared difference (low snowfall/500–1,000 m elevation)
= 41 – 34.209 = 6.791^2 = 46.118

This result, however, is dependent on the number of observations. The final calculation of the chi-square test standardizes the calculation regardless of the number of observations.

The sum of the cell values, 42.206, is the final chi-square statistic. We now need to consider how the number of variables influences the null hypothesis acceptance or rejection, something called "degrees of freedom" in statistics. You have already seen how the number of observations could effect the statistic and was taken account of. The number of variables can have an effect as well. The degrees of freedom is calculated by multiplying the number of rows—1—by the number of columns—1—in the observed or expected tables.

Degrees of freedom (df) = (3 – 1) × (3 – 1) = 4

Using a probability table showing degrees of confidence and confidence level (alpha), we can establish the number of times out of 100 that would be exceeded if the null hypothesis were true. In this case, with four degrees of freedom and at a confidence level of 0.05, the probability is 9.49. The value from the chi-square (42.206) is significantly higher. This means that the probability of getting a chi-square value this high is much lower. In other words, we can now pretty certainly say that we should reject the null hypothesis and note that the relationship between the amount of snowfall and elevation has been statistically supported.

From this point, we could move on to other statistical tests to look at specific relationships between snowfall and elevation. We might also want to assess our chi-square statistic, comparing it with other data and trying to make our statistic more robust. For example, we only used 20 observations in high elevation areas. Could we get more observations? Should we find out where the observations are located to determine if there are any biases? These questions reflect concerns that need to be expressed when using this data, as the chi-square statistic itself fails to account for general or specific geographical relationships.

Spatial Interpolation

A problem for many applications of geostatistics is that data is available for selected points in an area, but not for the entire area. A common example is soil pH, which can be collected only at distinct points using testing equipment. Basically, any ground, water, or air property is based on measurements recorded at individual points, making spatial interpolation a very useful transformation.

A number of factors influence spatial interpolation. Most significantly, the choice of technique will have great impacts. A number of statistical techniques, from simple to complex, are available. Choosing one depends on a number of factors and experiences. The two techniques that are often used are the local spatial average and the inverse-distance-weighted spatial average. Splines and multiquadric analysis and TINs are also widely used, but this section focuses on spatial averaging as a start.

Local spatial average interpolation considers multiple sample point values when determining the interpolation. The interpolation can specify the maximum area for calculating the average, the number of points, and the maximum distance a point can be considered in the average. Closely and regularly spaced sample points can be interpolated with little overlap or blank spaces. If the sample points are irregularly spaced or far apart, the interpolation may end up with gaps and deviate from the actual values considerably. Increasing the area for averaging may help, but can also turn the interpolation into averages that bear little resemblance to the actual area and the sample points. Just considering the nearest-neighbor sample points is more reliable, but can produce inaccurate interpolations if the points are far apart.

The inverse-distance-weighted spatial average interpolation offers some possibilities to address the weaknesses of local spatial average interpolation. This technique gives nearby points more significance in calculating the interpolation than more distant points. The weight is proportional to the inverse distance between the origin point and the sample point to be interpolated. In other words, as distance decreases the significance of the sample point increases. This weighting takes account of autocorrelation.

The weights (see Figure 14.6) vary by distance. The distance weighting can be altered. In the figure, it is close to a linear relationship between the distance and the weight. Other than the weighting itself, the resolution of the grid being produced by the interpolation and the order of choosing

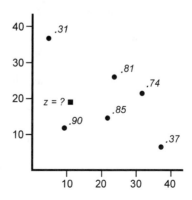

FIGURE 14.6. Example of weights assigned to elevation (z) based on distance point being evaluated.

points for evaluation can have subtle effects on the results. More specific control is offered by kriging.

Summary

Geostatistics offers the most complex and important GI analysis operations because of the underlying power of its mathematically orientated geographic representations. In geostatics the patterns recorded as GI represent spatial processes. The mathematical geographic representation is well suited for analysis of the spatial relationships among complex and diverse indicators that are beyond direct human interpretation—for example, the analysis of election results from the over 3,200 U.S. counties. Geostatistics usually works directly with measurements instead of with cartographic representations. Work with geostatics therefore requires paying attention to the complexity of establishing, creating, and validating measurements.

Because of the effects that aggregation to different units—for example, counties or states—has on measurements, spatial autocorrelation is an important issue to keep in mind. Basically, the principle of spatial correlation is that things are more similar to near things than to faraway things. Classical statistics assumes, however, that things, no matter where they are, are randomly related to other things.

The Modifiable Areal Unit Problem (MAUP) is the assumption that a relationship observed at one level of aggregation holds at another, more detailed, level—for example, that the majority of voters in a city voted for the candidate who won the state election simply because the candidate won the state election.

Geostatistics are applied in many areas for many purposes. Archaeologists doing initial surveys of an area or after collecting data rely on geostatistics to study possible relationships between data-use terrain analysis. Chi-square analysis is used as a basic technique to look at the strength of possible relationships, but is prone to a number of problems arising from autocorrelation.

In-Depth Kriging

The arbitrariness of distance-weighting functions used in spatial interpolation can be addressed by specifying the general form of the function and using point sample data to determine the exact form. The polynomial equations make it possible to see clear trends in the data, rather than to fit a rigid structure to the data with inverse-distance-weighting interpolation.

Kriging is based on a theory of regionalized variables, which means that distinct neighborhoods have their own variables. This leads to the calculations being optimized for neighborhoods rather than for the entire area. Kriging uses inverse-distance-weighting interpolation for each neighborhood, which can also be varied in a neighborhood. This involves three steps: (1) assessing the spatial variation of the sample points, (2) summarizing the spatial variation, and (3) using the spatial variation model to determine interpolation weights. The mathematics are summarized by a number of authors. Plate 13 shows an example of kriging.

Review Questions

1. What is spatial autocorrelation?
2. What is the "ecological fallacy"?
3. Why aren't geographical things and events random?
4. Why is statistical validity important? What are the two types?
5. How are geostatistics used in terrain analysis?
6. What parameters impact chi-square analysis?
7. Explain the Modifiable Areal Unit Problem.
8. What is the basic concept behind geostatistical analysis?
9. What is "spatial interpolation"?
10. What do patterns signify about processes in geostatistics?

Answers

1. What is spatial autocorrelation?

   Geographic data sets are neither random nor samples; as Waldo Tobler stated, "near things are more related than distant things." Any geographic data will be affected by its location.

2. What is the "ecological fallacy"?

   The ecological fallacy refers to the assumption that the statistical relationship observed at one level of aggregation holds at a more detailed level.

3. Why aren't geographical things and events random?

   The characteristics and processes of things and events are affected by where they occur.

4. Why is statistical validity important? What are the two types?

   Statistical validity measures the reliability of data and analysis. The two types are internal and external validity. Internal validity considers comparisons between measurements made at the same place over time. External validity considers how well data compares to similar measures from other places, other instruments, other types of observations, and other ways of recording data.

5. How are geostatistics used in terrain analysis?

   Geostatics can be used in different ways for terrain analysis. A distinction can be made between the initial geostatistical analysis of terrain and analysis that evaluates hypotheses and ideas generated from existing observations.

6. What parameters impact chi-square analysis?

   The distance and placement of observations have considerable impacts on chi-square analysis.

7. Explain the Modifiable Areal Unit Problem.

The Modifiable Areal Unit Problem (MAUP) is a special instance of the eco-logical fallacy, that results when data collected at a more detailed level of aggregation (e.g., census blocks, counties, biotopes) are aggregated to less detailed levels (e.g., census tracts, states, watersheds). Results valid at one level of aggregation may not be valid at another level.

8. What is the basic concept behind geostatistical analysis?

That patterns indicate geographic processes.

9. What is "spatial interpolation"?

Determining characteristics for places or areas based on existing observa-tions.

10. What do patterns signify about processes in geostatistics?

Patterns indicate underlying processes and relationships in geostatistics.

Chapter Readings

Gould, P. R. (1970). Is *Stastix Inferens* the Geographical Name for a Wild Goose? *Eco-nomic Geography*, *46*, 439–448.

Miller, H. (2004). Tobler's First Law and Spatial Analysis. *Annals of the Association of American Geographers*, *94*(2), 284–289.

O'Sullivan, D., & D. J. Unwin. (2003). *Geographic Information Analysis*. New York: Wiley.

Tobler, W. (1968). Transformations. In J. D. Nystuen (Ed.), *The Philosophy of Maps*. Ann Arbor: University of Michigan.

Web Resources

An illustrated introduction to the application of basic geostatistical analysis in archae-ology can be found online at *http://rst.gsfc.nasa.gov/Sect15/Sect15_10.html.*

A GIS-based terrain analysis exercise can be found online at *http://rockyweb.cr.usgs.gov/public/outreach/gislessons/firetower_dem_dlg_analysis_lesson.pdf*

For more archaeological examples, see *www.archaeophysics.com*

A tutorial a real-life adventure in environmental decision making can be found online at *www.nwer.sandia.gov/sample/ftp/tutorial.pdf*

Exercise

1. EXTENDED EXERCISE Chi Square Analysis

Objective

Learn basic concepts of chi-square analysis and issues that influence geostatistical analysis.

Description

The chi-square analysis is a straightforward statistical technique to evaluate the validity of hypotheses. While it can get rather complicated for large tables, for simple two-dimensional tables it is still a simple statistical technique. The only thing you should bear in mind when working with chi-square analysis and geographic information that geographic information is rarely, if ever, follows a normal distribution.

Instructions

First you need to prepare a combined map of precipitation and forested areas in Isaland. The null hypothesis is that forested areas are more common in areas with higher rainfall.

Here is list of the steps:

Prepare combined map

Precipitation by High/Low

Forest by Forested/Nonforested

Prepare tabulation

Prepare observed frequencies and expected frequencies matrices

Determine chi-square statistic

Calculate Yule's Q

Answer questions

You will use two equations:

Chi-square

$$\chi^2 = \sum \frac{(ObservedFrequency - ExpectedFrequency)^2}{ExpectedFrequency}$$

Yule's Q

$$\frac{ad - bc}{ad + bc}$$

Preciptation

H	H	H	L	L
H	H	L	L	H
H	L	L	H	L
H	L	H	L	H
H	L	H	H	H

Forests

F	—	F	F	F
F	—	—	—	F
F	L	L	F	—
F	—	F	—	F
F	—	F	F	F

Tablulation of Characteristics
High/Low Preciptation,
Forested/nonforested

Grid#	Precip.	Forest

Combined

Observed frequencies			Expected frequencies		
	Forested	Nonforested		Forested	Nonforested
High precipitation			High precipitation		
Low precipitation			Low precipitation		

Chi-Square Calulations

1	2	3	4	5
Matrix cell	Observed frequencies	Expected frequencies	$(O - E)^2$	$(O - E)^2/E$
a				
b				
c				
d				
				Sum =

To determine the chi-square value we will use an interactive web application or a published chart.

Calculate Yule's Q

Use the observed frequencies. A value close to +1 indicates a strong positive relationship between the independent and the dependent variables.

Yule's Q = _____

a, b, c, d stand for cells in the observed frequency table

When you are finished, answer these questions.

Questions

1. How strong is the relationship between precipitation and forests in Minnesota?
2. What factors explain the values?
3. What other factors should be considered?
4. What problems do you see in the procedure for this exercise? Are the quadrants a good size for this analysis?

Chapter 15

Futures of GIS

As the computerized collection, processing, and embedding of geographic and cartographic representations increases, GI will become an increasingly integral part of even more activities. The last three decades have seen tremendous changes in geographic information systems (GIS), the Global Positioning System (GPS), remote sensing technologies, and other information technologies. The next three decades will certainly witness similar magnitudes of changes. Arguably, GIS already plays the central role as the integrating platform for various technologies and organizations. GI will become as common as maps are today and have been for the last 100 years. Plate 14, an example of an interactive GI application with Google Earth, shows one of many exciting developments.

What the future means specifically for the technologies for use of, and access to GI is more difficult to say. It is almost certain that GI will become a commodity and that it will be produced, sold, and exchanged like other information commodities—information about businesses, for example. Costs and inequitable access to commercial GI may promote more government support for free or low-cost GI access, kick-starting a new vibrant market that develops applications and supports local democratic decision making. The ubiquity of GI also means increased surveillance by governments and private groups. For many, such unregulated surveillance is a considerable problem for civil liberties; for others, it is a necessary means to assure these liberties. A yet undecided key question in many places is how governments will allow access to their geographic information. The increasing prevalence of GI brings both opportunities and challenges.

To begin, we should note that GI is still underfunded. U.S. federal information technology spending is expected to be about $63.3 billion in 2007. In the same period, the 50 U.S. states are expected to spend $23.7 billion on information technologies. Local information technology spending in the United States in 2007 is expected to be somewhat higher: $27.2 billion. The entire GIS industry had revenues in 2006, according to Daratech, of $3.6 billion worldwide. Although government spending is only part of the spending on GIS, and although comparing spending to revenue is a crude measure, a comparison of the GIS revenue and U.S. total government spending suggests that GIS makes up only a tiny percent of all U.S. government spending. GIS is important, but most government dollars are going toward other information technologies. This becomes clear whenever a disaster strikes and the responding agencies are unable to coordinate their GIS because of conflicting standards and organizational responsibilities.

Even how government spending will change is a question for fortunetellers. Too many variables influence how government resources are spent. The above considerations also leave out the important roles of the GIS industry and the private sector. Instead of attempting to prophesize developments, the sections in this chapter focus on some of the questions underlying developments of GI and cartography in the next few years:

- *Where has GIS been?*
- *Where is GIS going?*
- *What are the ethical issues?*
- *Who pays for the data?*
- *What are the opportunities?*
- *What is the employment outlook?*

Where Has GIS Been?

Any history of GIS is a partial history. The technologies, institutions, and, most of all, the people are still under examination. We lack information about a number of key details. A number of authors (see the Chapter Readings) have offered various studies and stories about how GIS developed. Clearly, the deepest roots of what we call "GIS" today are complex and weblike. They reach back into developments before and especially during World War II of the first information technologies and computing technologies; they fit into a scientific penchant among government institutions to develop rational approaches to decision making and political desires to have empirical approaches to decision making, approaches that meld the strengths of information technology with vast repositories of statistical information that governments had begun to collect in the 19th century; they promote a scientific engagement with technologies that provide more robust modes of dealing with information; and they draw on geographers' and cartographers' work attempting to develop visualization techniques that were flexible enough to meet the needs of governments and private industries and readily taught at various levels to the technicians, analysts, and instructors needed to support these needs.

If we need a creation story, GIS starts out in the desire to harness information technology to a multitude of needs for representing geographic things and events, primarily with an emphasis on supporting government decision making. Its first proponents and developers came from urban research (Horwood), planning (McHarg), analysis (Fisher), and natural resource management (Tomlinson). Edgar Horwood was interested in examining how urban areas developed and was one of the first individuals to offer an academic course on the processing of GI. In 1963 he founded the Urban and Regional Information Systems Association (URISA), which still exists today. Ian McHarg, a landscape architect, became intrigued by the potential for improving and democratizing planning studies and published his seminal *Design with Nature* in 1968. Howard Fisher had begun working around the same time as Horwood and went to Harvard University in the mid-1960s to develop general-purpose automated mapping and analysis software. Under his direction, the Harvard Laboratory for Computer Graphics and Spatial Analysis became one of the key locations for developing early GIS applications and approaches. In parallel, Roger Tomlinson, who was involved in studies of land use in Canada during the late 1950s through an aerial survey company, began working closely with IBM and developed a system for managing land use information. In 1963 the team working on the project named this the "Canadian Geographical Information System."

Of course, many others were involved. Many other stories are crucial to the myriad ways GIS developed and became such a vibrant and significant technology. Certainly, the many interactions, formal and informal, that took place during the 1950s and 1960s were crucial to the development of GIS as we know it today. Jack Dangermond, a student of landscape architecture at Harvard University, heard about work in Fisher's lab. Grasping the opportunity, he founded a company in Redlands, California, that drew on lessons from the Harvard lab and has since become one of the most successful GIS software and consulting companies, the Environmental Systems Research Institute, or ESRI.

GIS for many years was a domain dominated by large companies and government agencies, especially agencies involved in mapping, such as the U.S. Geological Service (USGS), but the rapid fall of prices for computing hardware in the 1980s and 1990s led to many smaller government agencies and private companies becoming very significant. ESRI has undoubtedly been dominant in many areas; smaller companies that fill specific niches have become increasingly prevalent in the late 1990s. A new group of smaller companies that parallel the development and diversification that accompanied the development of the Internet, has become increasingly significant. The importance of standards and specifications indicates that GIS has been changing and continues to change.

Where Is GIS Going?

Although the future will certainly see a continuation of GIS usage for projects involving geospatial distributions and existing application areas, new directions are already evident that may alter the ways by which GIS develops. Primarily, these new GIS futures are in embedded applications and spatial data infrastructures. Both of these developments rest on widely established collections of easily accessible GI in clearinghouses and expertise resting with the many private and public users of GIS.

For the next few years the development of GIS technologies will probably focus on location-based services (LBS). *Location-based services* can be defined as technologies that add geographical functions to other technologies. The most common example is GPS-equipped phones. Originally developed to satisfy emergency services' requirements, the addition of GPS to mobile phones has opened up possibilities for cellphone companies to develop directory, mapping, and query services for consumers.

Right now, most LBS applications require programming or licensing software that is added to a mobile phone or other hand-held device. The next step in the eyes of many is the embedding of LBS with the functionality of GIS in mobile and household devices (e.g., navigation aids that not only find the theoretically shortest route from the store to your home, but also take traffic into account and adjust the route accordingly). (See Plate 15 for another interactive GIS application.)

The availability of such services depends on access to GI. Providing and maintaining the levels of institutional support required for these types of applications is very demanding and very expensive. A spatial data infrastructure (SDI) has become the way of thinking about aligning existing institutions with these developing GIS applications. SDI developments have faced enormous challenges. Costs are great, retooling existing institutions is always difficult, and finding political support has been challenging. A SDI-based approach seems the best strategy at the moment, but the difficulties have led to the creation of application-orientated data providers that circumvent the complexities of SDI and provide an easy-to-use and consistent system. The development of standards for web mapping have been crucial to these changes.

Indeed, the changes may be so significant that GIS begins to be recognized by other names, or even disappears. What a mobile phone user encounters when using a service to find nearby Italian restaurants has for that person little or even nothing to do with GIS. If the GIS is critical behind the scenes, it may only be relevant to a small group involved in the service's development. Improved ease of use may bring about a change in how we think about GIS.

What Are the Ethical Issues?

An important starting point for confronting any challenges related to geographic information is ethics, or the consideration of the principles of good conduct and how we should act. Ethics is different from morals, which are concerned with right and wrong and good and evil. GIS and mapping lead to many complicated ethical questions. These focus on locational privacy, surveillance, the collection and reuse of data, and the responsibility of professional groups.

One of the most important ethical issues for the future of GIS is locational privacy. Location-based services open the door to many intriguing and helpful capabilities, but also offer malicious people and groups opportunities to collect information about peoples' movements and offer companies possibilities to use the information collected for other purposes. The potential for abuse is so substantial that the infringment on locational privacy has been referred to as "geoslavery" by some well-know GIS Science figures.

Locational privacy is an issue that dovetails with concerns about increasing surveillance. The exaggerated powers of surveillance technology as shown in popular U.S. movies may be far off, but dense networks of surveillance certainly already exist. No vehicle enters the center of London without its license plate number being recorded. The surveillance networks of communist East Germany showed how much information can be collected even with the help of very modest information technology. The increase in the number of surveillance cameras coupled with data from mobile phones and GPS devices and other sensors opens the doors to unimaginable levels of surveillance.

The biggest constraint for surveillance remains the amount of data collected. The collection and reuse of data poses special challenges. Data collected legitimately for one purpose (e.g., the use of prepaid public transportation stored-value cards) can be linked to other data (e.g., images from a surveillance camera) to compile information on individuals that impinges on their privacy. Right now there are several legal limits in the United States on what private companies can do with information after it has been collected, but no blanket law defines privacy and its protection. How legitimate reuse of data should be regulated persists in posing special challenges.

Professional groups recognize a responsibility for establishing principles to help guide their members' activities through these very complex issues. The ethical guidelines or rules presented by various professional agencies provide a useful starting point for considering what is appropriate behavior in a variety of challenging situations.

Educators are also called upon to engage these challenges in their instruction. Generally, ethical issues fall short in technically orientated programs and even in professional programs compared to technological issues. The possibility is certainly there to engage ethical issues in conjunction with technical and organizational questions, which seems a more vibrant way to deal with the many changes facing GIS and mapping.

Who Pays for the Data?

Among these changes are substantial institutional changes related to the collection, maintenance, and publication of GI and maps. Large mapping organizations are challenged both internally because of the reduced size of government and externally by the increase in the number of service providers offering GI resources and services to key customer groups. The availability of low-cost GPS equipment has already greatly changed the way that maps for many tourist regions around the world are prepared and sold. Organizations (e.g., OpenStreeMap.org) promote the development of freely accessible geographic information using public domain standards.

One common approach to dealing with these changes has been for organizations investing in GI to attempt to recover all or some of the costs associated with gathering the GI. In some instances, public organizations even attempt to make a profit on the sales. While cost recovery has a few successes, for the most part more examples can be found where it has not worked. The reason for the failures has been often that the data is not necessary for the purposes, alternative data could be found, or people protested against the perceived overcharging for data.

Technology changes are fueling new approaches to cost recovery that may be more successful. Instead of relatively simple models for charging fees or licenses based on the amount of data requested, service-orientated approaches charge only for the data that has been used. These approaches are certainly the result of distributed geoprocessing, which makes it possible to access only the GI needed for a specific application over the Internet from as many computers, no matter where located, if needed.

Distributed geoprocessing needs special coordination of the underlying infrastructure. Internet connections break frequently, and if for some reason this happens unbeknownst to the computer operator, a serious problem can result. This also applies to the information technology used for displaying and working with the GI. Distributed data, as a result, may be incomplete in subtle but important ways. Distributed geoprocessing may not work in many circumstances, no matter the perceived advantages. This is one of the main limitations to these approaches, and as such is very technical in nature, requiring special technical services and/or precautions. Another challenge, along these lines, the reliable provision of Internet-based services over wireless networks and through cables, opens the doors to many opportunities for GIS and mapping.

What Are the Opportunities?

The new and ongoing opportunities in GIS are so many that a separate book on them would be necessary to do them justice. Certainly, advances in technology mean that LBS-based approaches have received a great amount of attention and will continue to be a focus for developments in the future. The underlying technology requires technically adept individuals; the applications require many more individuals that can make the technology work and best support various needs.

Employment

The opportunities for employment range from research and development to support. On the one end, we find the people involved in creating and developing the technologies and devices. They are usually engineers, but opportunities in this domain also exist for people who work on testing and refining devices and programmers and analysts who play key roles in development. Support generally involves people who help users work with the technologies and devices as well as through training. In between, many people work on developing and maintaining applications either as programmers or as specialists for particular application domains. Last, but by no means least important, people working in marketing are often involved in assessing consumer demands and helping companies identify profitable areas to work in. It is certainly possible to find individuals who work in all domains during the course of their normal work week—the needs of smaller companies often require such talents.

Opportunities can be found in all sectors. Government positions abound in all areas, but are especially important for applications and to a lesser extent in research and development and in support. Private industry offers a large number of applications as well, and has the most positions in research and development. Nongovernment and nonprofit positions are scarcer, but many people find that they offer a better balance between quality-of-life and income than jobs in other sectors.

Summary

GIS has experienced a dynamic past and adapted to the advent of new technologies. Currently GIS seems set on at least partially morphing into more visualization capabilities focused on specific applications. LBS appear to be a key driving force for the next round of GIS development. Opportunities abound in this area, but also continue in traditional GIS areas. These developments open ethical issues. The capability of tracking people through the entire day is now readily possible, but how should we know and control how we are tracked and placed under surveillance? Challenges for the development of GIS touch on these issues, but other fundamental technical and organizational challenges remain.

In-Depth What Is a Spatial Data Infrastructure?

There are many definitions of what makes up a spatial data infrastructure (SDI). Common to all is that an SDI facilitates and helps coordinate the exchange of spatial data. This can be restricted to stakeholders, or limited to a defined community. Sometimes the limits are not crystal-clear, but stakeholders and community members find better services, data, and support than others.

Problems arise because any SDI is understood in different ways by its stakeholders and community members. SDIs seem to have been most successful when they support decision making and facilitate interactions between organizations.

FIGURE 15.1. GPS serves a wide range of urban applications.
MobileMapper™ CE image courtesy of Magellan Navigation Inc.

FIGURE 15.2. GPS also serves a wide range of agricultural and environmental applications.
MobileMapper™ CE image courtesy of Magellan Navigation Inc.

Review Questions

1. Do you think that geographic information will become pervasive and ubiquitous? Why or why not?
2. How significant are costs for the development of geographic information?
3. Why is government funding of GIS in the United States so small in comparison to overall funding of information technology?
4. GIS's history involves many fields. What are some of the most important fields?
5. What is the name of one of the most important U.S. federal government agencies involved in GIS?
6. What does "LBS" stand for?
7. What is an SDI?
8. What are important ethical issues for GIS?
9. How are changes to technology impacting cost-recovery models of GI pricing?
10. What is the range of employment opportunities?

Answers

1. Do you think that geographic information will become pervasive and ubiquitous? Why or why not?

   There are no right or wrong answers, just thoughtful and unthoughtful ones.

2. How significant are costs for the development of geographic information?

   Costs are one, if not the most, significant issue, for the development of geographic information systems, no matter what form they take.

3. Why is government funding of GIS in the United States so small in comparison to overall funding of information technology?

   GIS is only one of the technologies used by government. Major investments in health, welfare, public safety, and many other areas do not involve computing.

4. GIS's history involves many fields. What are some of the most important fields?

   Landscape architecture, natural resource management, planning, geography, computer science, and cartography.

5. What is the name of one of the most important U.S. government agencies involved in GIS?

   The United States Geological Survey (USGS)

6. What does "LBS" stand for?

   Location-based services

7. What is an SDI?

The alignment of various institutional GIS to support multiple GIS applications and users, some of whom may be unknown and undefined, but important in the future.

8. What are important ethical issues for GIS?

Many questions: What do people do with data? How is the data collected? How will locational privacy be protected? How to charge for geographic information?

9. How are changes to technology impacting cost-recovery models of GI pricing?

GI is becoming a resources for services, used as needed by users and charged for according to use.

10. What is the range of employment opportunities?

Research, application development, and support.

Chapter Readings

Chrisman, N. (2006). *Charting the Unknown*. Redlands, CA: ESRI Press.

Curry, M. (1998). *Digital Places: Living with Geographic Information Technologies*. New York: Routledge.

Fisher, P. F., & D. J. Unwin (Eds.). (2005). *Re-Presenting GIS*. Chichester, UK: Wiley.

Foresman, T. (Ed.). (1998). *The History of Geographic Information Systems: Perspectives from the Pioneers*. Upper Saddle River, NJ: Prentice Hall.

Pickles, J. (2004). *A History of Spaces: Cartographic Reason, Mapping, and the Geocoded World*. New York: Routledge.

Sheppard, E. (1993). Automated Geography: What Kind of Geography for What Kind of Society? *The Professional Geographer, 45*(4), 457–460.

Web Resources

OpenStreetMap is a free editable map of the whole world: *http://openstreetmap.org/*

The U.S. National Academy of Sciences, Mapping Sciences Committee, periodically publishes reviews and perspectives. The latest is *Beyond Mapping* available online at: *http://darwin.nap.edu/execsumm_pdf/11687.pdf*

The GISCI website provides a concise list of important principles as well as discussion of ethical issues for GIS professionals. See *www.gisci.org*

Typical for what many organizations do to have and maintain GIS, saving and scavenging: *www.dailymail.com/story/News/+/2006083137/Frustrated+county+officials+to+use+local+funds+for+mapping/*

Google Earth is now famous for its mapping and visualization capabilities. See *http://earth.google.com*

Index